Electronic Power Control

Electronic Power Control

Irving M. Gottlieb

TAB BOOKS
Blue Ridge Summit, PA

Notices

Alinco™	Magnetics and Electronics Inc.
Dow Corning®	Dow Corning Corp.
Fair-Rite™	Fair-Rite Products Corp.
HEXFET™	International Rectifier Corp.
Radio Shack®	Tandy Corp.
Sprague™ **(capacitors)**	Ico Rally Corp.
Sprague™ **(electronic components)**	Sprague Electric Co.
Variac™	Penril Corp.

FIRST EDITION
FIRST PRINTING

© 1991 by **TAB Books**.
TAB Books is a division of McGraw-Hill, Inc.

Library of Congress Cataloging-in-Publication Data

Gottlieb, Irving M.
    Electronic power control / by Irving M. Gottlieb.
      p. cm.
    Includes index.
    ISBN 0-8306-2453-8   ISBN 0-8306-2445-7 (pbk.)
    1. Power electronics. 2. Electronic control. I. Title.
TK7881.15.G66 1991   91-4712
621.31′7—dc20   CIP

TAB Books offers software for sale. For information and a catalog, please contact TAB Software Department, Blue Ridge Summit, PA 17294-0850.

Acquisitions Editor: Roland S. Phelps
Book Editor: B.J. Peterson
Production: Katherine G. Brown

Contents

Preface

THE WONDERFUL WORLD OF SOLID-STATE POWER CONTROL BEGAN WITH the germanium power transistor and the SCR (silicon-controlled rectifier). These devices quickly made extensive inroads into power applications that had previously depended upon vacuum tubes, thyratrons, magnetic amplifiers, and electromagnetic relays. This is hardly surprising because a number of compelling advantages were realized: smaller and lighter packages, relative freedom from mechanical and other wear mechanisms, more manageable and convenient operating voltages, extended frequency response, and enhanced reliability. These features offer cost savings as well.

Because of current leakage and excessive temperature dependency, germanium devices gave way to silicon devices. Silicon devices, in turn, have undergone continuous progress in power, voltage, current, and frequency capabilities, and in other parameters. SCRs have evolved considerably from their early versions. Originally relegated to 60 Hz (Hertz) service, these devices developed to the stage where kilowatts of power could be controlled at many kilohertz. Also, the triac entered the scene, allowing full-wave control of ac power. For a decade or so before 1975, triacs remained the state of the art for power-control devices. It appeared that this technology had attained maturity, and that future progress would be limited to ever-diminishing refinements.

Beginning in the mid-1970s, and gathering steam during the 1980s, a new trend in power devices and control systems became identifiable. Even though the bipolar power transistor had been marvelously improved, other devices began to assume prominence. Notably, the MOSFET (metal-oxide-semiconductor field-effect transistor) shed its former lightweight status, becoming a heavyweight contender in power control. In somewhat similar fashion, the Darlington device invaded power levels and frequency domains previously thought unattainable.

With these developments came awareness that sophisticated ICs (integrated circuits) could greatly simplify implementation of the complex logic and driver circuitry often needed in conjunction with power devices. This advancement, a notable achievement in itself, led also to the fabrication of logic, driver, and power stage on a single chip or within a module. This integration probably suggests the nature of future progress in power-control techniques.

Recent and emerging power devices might use a number of constructions and semiconductor formats; these could range from discretes to "smart" systems. There are power ICs, MOS-bipolar integrations, power device-control logic combinations, hybrid modules uniting the best in power devices and analog or digital ICs, thyristors with self-contained trigger logic, optoelectronic devices with respectable power capability, and other implementations intended to reduce interconnections and parts counts.

The discussions and application examples in this book should prove rewarding to the practitioner motivated to keep up with present and anticipated trends in the electronic control of power. For balance, a good measure of space is allotted to many basic techniques, which because of their time-proven usefulness are unlikely to be soon obsolete. The format of this book caters to those who like to build and experiment with their own circuits. Except for the first chapter, which is introductory, you will find the presentations replete with component values, parts lists, and practical guidance.

Introduction

THE BEST USE OF THIS BOOK WILL DEPEND TO A CONSIDERABLE EXTENT on awareness of its intended purpose. First and foremost, it is the author's objective to make *Electronic Power Control* a practical guide to serve the interests and needs of a wide range of electronics practitioners.

In order to accomplish this goal, care has been taken to explain principles of operation in descriptive, rather than mathematical, language to the greatest degree. At the same time, the overly simplified approach has been avoided. The author assumes you are familiar with basic electrical and electronic fundamentals and have some practical experience. With this minimal background, the book is intended to benefit hobbyists, experimenters, hams, technicians, service people, engineers, and physicists.

Readers will find it convenient to refer quickly to the section of the book catering to the application of his or her interest. Please review chapter 1 first; then, it will not be necessary to read the rest of the chapters sequentially.

This book is not intended to be used as a how-to treatise. It is anticipated that the majority of applications will be unique and will have individual requirements geared to conditions at hand. Accordingly, readers will wish to scale power levels up or down, limit or expand circuit capabilities, or otherwise modify the various circuits. In this feature, the practical-guidance aspect of the book will be valuable. The majority of circuits are enhanced by inclusion of component sizes or parts lists. The basic sections of systems are explained in a way that facilitates modification.

The author recommends that, regardless of the individual application needs, you first make circuits operational as depicted in the book. Then you can make careful modifications to optimize the circuit or system usefulness to the unique demands of a particular application. This approach will be the best for analysis, troubleshooting, and practical success.

1
Overview and generalities

THIS CHAPTER PROVIDES USEFUL GUIDANCE IN ELECTRONIC CONTROL OF power. It seems likely that two main classes of readers can profit from the material presented here. Those with basic electronic know how, but without a specialization in power control, will find valuable insights that will facilitate practical implementation of the circuits in the subsequent chapters. The professional designer who might have forgotten more than the author knows will welcome the quick refresher of fundamental facts pertaining to the control of power.

Even those practitioners with very high proficiency in the art all too often tend to lose awareness of the price paid for all of the wonderful features of solid-state devices. Although they are compact, efficient, and low cost, these devices are by no means as forgiving as were electro-mechanical devices, or even the electron tube of yesteryear. The cherry-red anode of a power tube did not necessarily signify catastrophic destruction, or even an unacceptable shortening of life span. A glowing PN junction, on the other hand, reverts the silicon power device to silicon rock. Other than obvious thermal overload, many bugs lie in wait for the opportunity to perpetrate a lethal invasion. Included are power-line transients, excessive rate of change of load current or of load voltage, unplanned current surges at cold start of incandescent lamps and motors, and parasitic oscillations from circuit and circuit-board elements.

The bottom line is that solid-state power devices operate in a non-ideal world. Often, mathematical analysis does not suffice for reliable results unless a good measure of intuitive judgement is also thrown in. The philosophy of over design can be both rewarding and costly. Over design is usually not as simple as with passive devices; it is easy enough to ensure reliability of a filter capacitor by specifying double the highest voltage rating it will see. Active devices, however, can't always be handled in this fashion because of the numerous trade offs often involved. If you specify a larger transistor than appears to be needed, you must determine that frequency capability, current gain, or leakage current did not adversely trade for the sought increase in allowable power dissipation. Nor should it be overlooked that parameter tolerance in solid-state power devices can span a range for a given device equivalent to two or three electron tubes of a different designation.

The mundane and the sophisticated

The exploration of power-control devices and techniques deals with the tried and proven as well as with the new. It is not always easy to identify the difference between ordinary power devices and those that might do justice to such labels as sophisticated, exotic, or smart. Most bipolar power transistors of the types that have faithfully performed in myriads of power circuits, such as regulators, amplifiers, and motor control, are ordinary traditional devices. Admittedly, some bipolar power transistors with inordinate switching speed, frequency capability, or current or voltage ratings are certainly not garden-variety devices. Perhaps a reasonable test is to judge in terms of performance generally achieved five or ten years ago. If there has been a definite advance in certain parameters of performance, the device might well qualify as an extraordinary one; (that is, one that is of special interest for modern designs) and as a possible forerunner of future development.

Interest in the newer and smarter power devices does not necessarily imply that the device must be integrated or associated with a microprocessor. Generally, there will be one or more plus features over devices commonly used a few years ago. For example, the modern power Darlington produces power gains not readily realized in the elemental power transistor. And, power MOSFETs require so little drive that they certainly merit classification as a modern power device destined for greatly expanded application in future designs of control circuits and systems. Then, too, power op amps and power ICs can do things beyond the capabilities of ordinary power transistors, unless assisted by auxiliary devices and circuitry. Even more suggestive of future trends are the now-developing families of dedicated devices and those with self-contained logic or drive circuitry.

The ultimate development of such smart power devices is a complete system or subsystem within a module. Figure 1-1 depicts the basic idea—instead of a collection of discrete devices, stages, and associated passive components. All or most of the power-control function is accomplished within a single package. Not shown is internal protection against overload, transients, and temperature rise. This concept is ideal. There are thermal, electrical, fabrication, and cost factors that mitigate against universal realization of such an elegant format.

Power versus energy

If you investigate techniques for controlling power, it makes practical sense to know what is being manipulated. In the first place, although your interest centers on electrical or electronic control of power, the power actually being controlled can manifest itself electrically, mechanically, thermally, chemically, or in other ways.

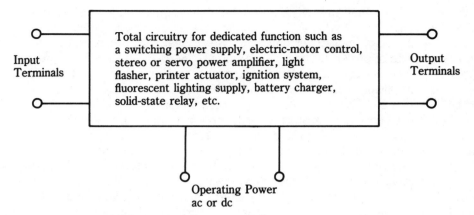

1-1 The ultimate power-control device—a system or subsystem on a chip or within a module.

Thus, power must have a common aspect applicable to diverse situations. It is often the case that a person who talks glibly about a powerful motor might not be altogether clear in defining this capability of the motor or other device. What, indeed, is *power*?

At the outset, realize that *power* expresses a rate of doing something; specifically, power is the rate of using, transforming, or dissipating energy. What is energy? *Energy*, despite its many and varied manifestations, is the capacity or ability to do work. Finally, *work* in its elemental form always involves displacement of a force through a distance. Obviously, instead of providing a quick and easy definition of power, one entity has just been telescoped into another, resulting in complexity rather than simplicity. It turns out, however, that it is acceptable to say that energy is work, and that power is the rate of doing work, or the rate of utilizing energy. Thus, a powerful motor is one capable of doing much work in a given time. Commonly used energy units are depicted in Table 1-1.

Note the appearance of the quantity, *time*, in the above discussion. Because power is described as a rate, it is inevitable that time is always an implied aspect of power. For example, 1 horsepower is an energy or work rate of 33,000 foot-pounds

Table 1-1. Conversion factors for commonly used energy units.

	Kilowatt-hours	Joule	Foot-pounds	BTUs (British Thermal Units)
One kilowatt-hour	1	3.6×10^6	2.655×10^6	3413
One joule or one watt-second	2.78×10^{-7}	1	0.738	9.48×10^{-4}
One erg	2.78×10^{-14}	10^{-7}	7.37×10^{-8}	9.48×10^{-11}
One foot-pound	3.77×10^{-7}	1.356	1	1.285×10^{-3}
One BTU	2.93×10^{-4}	1055	777.9	1

per minute. Note that all is visible here—distance, force, and time. Inasmuch as the product of distance and force represents work, or energy, 1 horsepower is a certain expenditure of energy per unit time. If you deal with other than mechanical manifestations of power, the factors making up energy might not be expressed in pounds or feet; but equivalent force, distance, or work units are involved. The net effect is that power expressed in electrical terms, such as watts, differs only from mechanical horsepower by a simple conversion factor.

Note that energy or work requires a force acting along a length, but *torque* requires the force to act at right angles to a length. The basic torque unit is the pound-foot. In contrast, the basic energy or work unit is the foot-pound.

From the previous discussion, you can see that the interchangeable use of the terms *power* and *energy* in the popular literature cannot be valid. You must always be aware that power is the rate given by energy per unit time. By the same token, if power is the known quantity, then energy is the product of that power and the time it is manifested. A practical example of this relationship is the watt-hour meter that obligingly provides the necessary billing information to the electric utility company. You are charged for the use of power over its relevant duration of time. The watt-hour meter simply multiplies power by its time of use in order to compute energy. You pay for energy because it does not matter to the utility company whether you consume a great deal of power for a short time or low power for a proportionately longer time. Commonly used power units are shown in Table 1-2.

Faced with a choice of having the utility company make either high energy or high power available to you, which would be the better choice? Without certain assumptions, the answer is neither. High energy might only allow the burning of a small 25 W lamp continuously for a year. Thus, low power coupled with a long time could satisfy the promise of high energy. Conversely, high power could conceivably allow the operation of a refrigerator, washing machine, and dishwasher for a limited time, such as an hour.

You can appreciate that in most instances high power can be meaningless unless such a power level can be sustained. The high power you wish to enjoy from the utility company must be available around the clock. Although rarely stated, you generally assume this to be the case. In certain technologies, this obvious assumption might not hold. A rocket motor, for example, should generate high power within an acceptably short time. A reciprocal situation is a large storage battery that might require considerable energy to charge, but will do so from a low-power source if sufficient time is allowed. If the utility company is to impress you

Table 1-2. Conversion factors for commonly used power units.

	Watts	Kilowatts	Horsepower	Ft-lb/min	BTU/hour
One watt	1	1×10^{-3}	1.34×10^{-3}	44.25	3.413
One kilowatt	1×10^{-3}	1	1.34	4.25×10^4	3413
One horsepower	745.7	0.7457	1	3.3×10^4	2545
One ft-lb per minute	2.26×10^{-2}	2.26×10^{-5}	3.03×10^{-5}	1	7.713×10^{-2}
One BTU per hour	0.293	2.93×10^{-4}	3.929×10^{-4}	12.97	1

with its high energy claim, this must be accompanied by an acceptable rate of delivery of the energy, which means a power rating. As stated, you must be satisfied that the desired power capability is on a sustained basis. The essence of this discussion is that power and energy are not interchangeable terms despite their mutual involvement in producing forces, displacements, and work.

Energylike parameters

The foregoing discussion of power and energy relates in an interesting and often practical way to matters encountered in electronics technology. Consider, for instance, the ratings for fuses intended for protection of power semiconductors. Whereas industrial-type fuses are selected according to their current ratings, semiconductor-protection fuses also have I^2t designations often described as *energy let-through* ratings. This nomenclature can be confusing, however; I^2t is *not* dimensionally equivalent to energy. In actual practice, one consults the spec sheet for the I^2t rating of the semiconductor device to be protected. The I^2t rating of the fuse must then be less than that of the semiconductor device. The fuse also must be capable of carrying the normal current range drawn by the semiconductor device.

The basic idea is to ascertain that a fault condition will enable the fuse to blow before the semiconductor device can be damaged. Because of the involvement of time in this manner, it can be loosely stated the fuse prevents damage from excessive energy despite the fact that I^2t does not strictly represent energy. I^2t is a practical means of matching fuses to power devices.

Such fuses also can carry maximum voltage ratings just as industrial-type fuses do. The I^2t concept is useful in that a proper fuse can be specified without resorting to much empirical (and costly) testing.

Another energylike parameter is the volts-second values that must be taken into consideration when passing a duty-cycle modulated wave through a transformer. Volt-seconds are not dimensionally equivalent to energy, but this parameter can relate closely to both power and energy considerations. This is because the secondary voltage of the transformer is governed not only by the turns ratio but is affected by the duty cycle. The explanation of this often-overlooked phenomenon will be found in this chapter.

It is natural enough to associate power ratings with power devices, but in some instances energy ratings provide more meaningful information to the designer. For example, the protective capability of the MOV (metal-oxide varistor) is determined by its energy absorption rating—usually in joules. And, secondary breakdown in power transistors is a destructive mode that is essentially energy dependent. This is why narrow-pulse operation tends to yield a more favorable SOA (safe operating area of the load line) than dc or continuous operation.

Power and energy at the submicroscopic level

The photon is the particlelike entity responsible for the energy content of electromagnetic waves. The theory of the photon holds that its energy is directly proportional to its frequency. That is why X-ray radiation is said to be more energetic than ordinary visible light. This concept becomes more meaningful when exemplified in the following way. Consider two 50 W (watt) RF (radio frequency) power generators, one at 10 MHz, the other at 100 MHz. Ten times as many 10 MHz photons as 100 MHz photons must be generated to produce the power level of 50 W. That is, 10 MHz photons have only $1/10$ the energy content of 100 MHz photons. Although practical design and operation of RF equipment is ordinarily carried out without consideration of photonic principles, those working at higher frequencies involved in opto-voltaic devices, lasers, and fiberoptics often profit from consideration of the energy content of the photon.

A classic example of the need to differentiate between power and energy is Einstein's explanation of photoelectric emission of electrons. It was shown that below a certain frequency of light, no amount of power could provoke photoelectric emission. Conversely, above a certain high-frequency threshold of the illumination, the smallest measurable power level sufficed to induce a proportionately tiny photocurrent. This simply stated concept merited the Nobel prize. If, however, sloppy use has been made of the terms *energy* and *power*, the idea would have been meaningless. The setups of Fig. 1-2 demonstrate that energy and power are not inter-

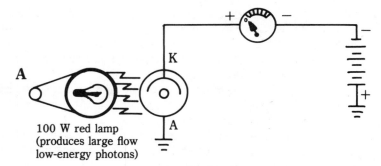

High-power red lamp produces no photocurrent.

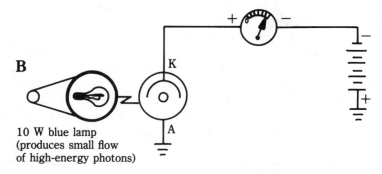

Energetic photons from low-power blue lamp produces photocurrent.

1-2   Phototube circuit demonstrates difference between power and energy.

changeable terms. Blue light is higher in frequency than red light. A certain threshold energy is needed to liberate electrons from the photoemissive cathode. Blue-light photons possess this energy; red-light photons do not, and no matter how many of them are involved (high power), no photoelectrons can be produced.

The quest for transporting energy in zero time

For most mundane purposes, it is convenient to assume that energy can be conveyed instantaneously from point to point. A corollary of this is the notion that the control of power at a distant load can also be accomplished in zero time. Pioneers in electrical communication found, however, that telegraph lines and submarine cables, because of distributed inductance and capacitance, always imparted delay times between initiation and arrival of signal information. But, even in a hypothetical line devoid of all reactive effects, you could not expect to send energy, control power, or transmit intelligence faster than c, the speed of light in free space.

Until the arrival of the space age, this restriction of nature was not of great importance to others than theoreticians and mathematicians who devised elegant relationships using c to optimize the performance of wire lines so that distortion and attenuation of signals were minimized. Now, however, it would be a blessing many times over if greater than c transit of energy could be achieved over interplanetary and interstellar space. To the best of our knowledge, nature always imposes insurmountable obstacles preventing this from being done. The following is a description of a down-to-earth situation exemplifying the limiting action of the speed of light, c.

Figure 1-3 depicts a resistive load receiving microwave power through a waveguide. In (A) the frequency (and therefore, the wavelength) of the energy impressed at the input of the waveguide is varied. This enables a plot of two curves to be made. One is designated as group velocity, V_g. The other is phase velocity, V_p. Both of these entities are wavelength or frequency dependent. The tantalizing thing about phase velocity is that it is greater than c, and, indeed, approaches infinity as the wavelength is made greater. Having discovered something capable of travelling faster than c, why not use it to control power in distant spacecraft and establish minimal-delay communications with it?

This is, indeed, a tantalizing question. Phase velocity V_p can be determined by measuring the wavelength λ_p in the guide by means of a probe moved along the slot. Then, $V_p = \lambda_p \times f$, where f is the frequency (which can be measured with a counter connected across the generator). Interestingly, f never changes within the guide. Note that wavelength times frequency is the general formula for determining the velocity of electromagnetic waves. In free space, it always yields value c.

The determination of V_p yields a velocity greater than the value c because λ_p inside the guide is always greater than the wavelength in free space. Why didn't you use some means of directly measuring phase-velocity V_p? That would, indeed, have been more satisfying, but it cannot be done; V_p is not a real parameter—you must be content to regard it as a mathematical entity useful in making calculations, but with not enough substance to grab hold of. This is a roundabout way of

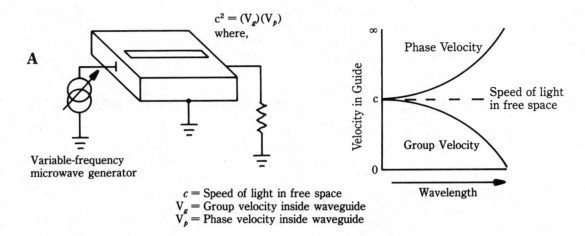

$$c^2 = (V_g)(V_p)$$
where,

A

Variable-frequency
microwave generator

$c =$ Speed of light in free space
$V_g =$ Group velocity inside waveguide
$V_p =$ Phase velocity inside waveguide

Within guide, phase velocity can exceed c but cannot convey energy or information to the load.

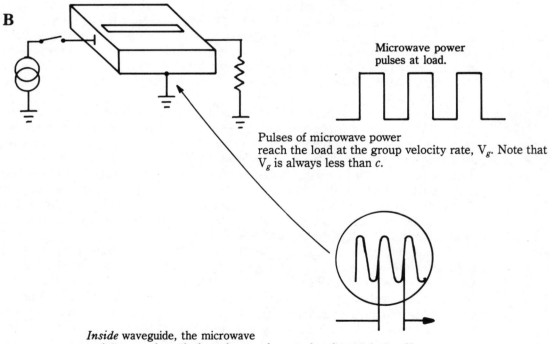

B

Microwave power
pulses at load.

Pulses of microwave power
reach the load at the group velocity rate, V_g. Note that
V_g is always less than c.

Inside waveguide, the microwave
cycles move through the pulse envelope at the phase velocity, V_p.

With pulsed microwave source, the pulse power reaches the load at a less-than-c rate. Within
the guide, the faster-than-c phase velocity of individual cycles moves right through the pulse
envelope.

1-3   Energy or information cannot travel faster than c.

saying that energy cannot be transported, power cannot be controlled, and information cannot be sent at rates exceeding the value c. (The last assertion makes a nice thesis for philosophical debate, but that must be considered outside the scope of this book.)

Having come this far, you can calculate the group velocity V_g from the relationship:

$$V_g = \frac{c^2}{V_p}$$

V_g is the velocity at which pulses, or other modulation formats, pass through the guide. It is the rate at which power can be controlled in the load. It is conceivable to measure directly the delay between the impinging and exiting power pulses at the waveguide. In any event, group velocity is always slower than c by the same percentage that phase velocity exceeds c.

In conclusion, this discussion has practical ramifications for those working with microwave power systems, but it cannot be used to mock nature's hard-and-fast rule about speed barrier, c. Those seeking faster or instantaneous rates of energy transport will have to discover some presently unknown waiver to this rule.

Switching transients— a hidden gremlin in power control

Making and breaking the flow of current in an electric circuit is the utmost in simplistic procedures . . . or is it? Suppose the current has been flowing for some time in Fig. 1-4A and the knife switch is suddenly opened. Because of the energy stored in the magnetic field of the inductor, you know that the abrupt cessation of current in the circuit does not come about in step with the physical breakage of contact between the switch elements. Several things happen before you can truly identify an open circuit. As the switch elements part, a high voltage develops across the switch blades many times greater than the battery voltage. This so-called *counter EMF* is accompanied by an arc dissipating energy in the form of heat, light, and sound. Additionally, considerable RF energy can manifest itself as interference in communications and other sensitive equipment. Once the arc has depleted the energy stored in the magnetic field of the inductor, it will extinguish itself, allowing the circuit to finally become open. A price might be paid for this process in the form of burned or fused switch blades.

Unfortunately, the same process tends to occur in a solid-state device used in switching applications. The semiconductor material inhibits the formation of the destructive arc up to a point. This feature enables the device to serve as a switch within the boundaries of its safe operating area. On the other hand, such devices are very unforgiving when excessive current is switched or considerable inductance is present. Various circuit techniques can be used to prevent destruction of the power switching-device by absorption of the excess energy. Circuits used for this purpose are generally referred to as *snubbers*, or *clamp* circuits. An important

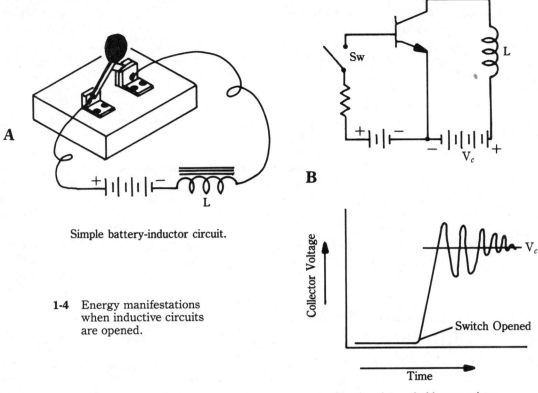

A

Simple battery-inductor circuit.

1-4   Energy manifestations
when inductive circuits
are opened.

B

Circuit using switching transistor.

aspect of the use of such protective circuits is that the stray inductance in a switching circuit can suffice to produce destructive counter EMFs in nanoseconds, even though there is no inductance in the load proper. This happens when high currents are being switched at high repetition rates, and with short turn-off times. Figure 1-4B shows the commonly encountered situation in solid-state switches.

The practical aspect is that the snubber circuit cannot be optimally effective unless it is connected to the power-switching device via very short leads. Otherwise the protected power switch can suffer catastrophic destruction; it is common that elegantly calculated snubber and clamp networks fail to absorb nanosecond energy excesses at the actual terminals of the switching device. This condition might not impart sudden death to the switching device, but it is often the cause of a mysteriously short life span.

When a bipolar power transistor is switched to its off conductive state, it is said to be *reverse biased*, although this might correspond to zero base-emitter voltage. Often, however, a reverse base-emitter voltage is actually applied during turn off. This shortens the storage time of minority carriers and reduces the fall time of turn off. Reduced fall time decreases switching losses because less time is spent during simultaneous collector voltage and collector current. You must be careful, however, for even though switching losses are lessened by reverse bias, the vulnerability to destruction from secondary breakdown generally increases. For practical purposes, keep in mind that secondary breakdown usually takes place at lower energy levels under reverse-bias operation (when the transistor is switched off) than when the transistor is forward biased (during the on state of the transistor). This situation is not desired because the switch-off interval is just when the inductive kick back of circuit inductance produces the dreaded switching transient. Commonly encountered switching waveforms are shown in Fig. 1-5.

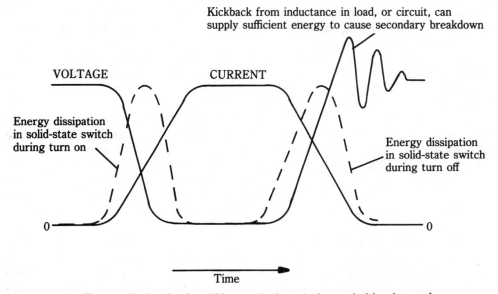

1-5   Energy dissipation in solid-state devices during switching intervals.

This book does not deal with rigorous methods for calculating snubber and clamping-circuit component values for optimum absorption of switching transients. The qualitative fact that these nonforgiving excursions of the load line exist, will suffice for most experimenters to take some practical precautions. The trouble with a formal mathematical approach is that the calculations often are not of the simple variety. Although the solution thereby attained might be elegant in cost effectiveness and in overall efficiency of the switching circuit, these need not be the primary goals when not designing for production runs. Information available about stray circuit parameters is usually not too closely in agreement with actual hardware. Even worse, manufacturer's data for power-switching transistors involves sloppy tolerances and often it is not easy to interpret by nonspecialists.

One precautionary procedure the hobbyist or experimenter can invoke is to use a switching transistor with greater voltage and current capability than appears to be needed. Do not go overboard in switching speed, for transistors with higher switching speeds invariably exhibit lower secondary breakdown energy levels. A transistor with just ample turn-on and turn-off times might dissipate more power during switching transitions, but it is likely to do so safely. Always select power transistors expressly intended for operation in switching circuits. These devices are made with as much electrical ruggedness as the art permits, especially with regard to reverse-bias secondary breakdown. At moderate switching rates, approximately 20 to 50 kHz, it is often wise to try to get by with just zero volts turn off at the base-emitter junction.

Compounding your troubles in effectively absorbing the energy in switching transients is the physics of the solid-state switching device. Obviously, you are not dealing with the simpler situation of the knife switch. It turns out that the time-honored concept of maintaining a low average device temperature via heat sinking does not in itself indicate safety from switching transients. Even worse, well-below safe junction-temperature rise will not necessarily prevent catastrophic destruction from the energy in switching transients. For the moment, confine your attention to bipolar devices—both discrete power transistors and power Darlingtons.

Characteristic curves of bipolar transistors show that a primary voltage breakdown occurs at a certain collector voltage, V_{cc}. This is basically an avalanche phenomenon and can readily be prevented from being destructive by limiting the collector current to maintain rated power dissipation. The energy in switching transients can cause another type of breakdown, appropriately known as secondary breakdown. The salient feature of secondary breakdown is the formation of hot spots within the device that do not materially contribute to the average temperature. Within their localized domain, however, these hot spots can melt the silicon or otherwise destroy the PN junctions. The common effect being an internal collector-emitter short.

There are two operating conditions giving rise to destruction via secondary breakdown. One is identified as FBSOA (forward-biased safe operating area). The other is identified as RBSOA (reverse-biased safe operating area). The SOA curves are graphical plots of collector current vs collector voltage; they show what values of collector current and voltage are permissible, that is safe when applied simultaneously under prescribed conditions of temperature, bias, and pulse duration. (See Fig. 1-6.) These curves will tell you that it is unsafe to apply both rated voltage and rated current at the same time. Furthermore, the SOA curves tell you that heed must be paid not only to limits imposed by current, voltage, and power dissipation, but by secondary-breakdown energy constraints. Inasmuch as secondary breakdown is energy dependent, it should come as no surprise that secondary breakdown can occur at a lower collector voltage than that attributed to primary breakdown. In any event, the load line must not penetrate the SOA curves pertinent to the pulse duration, reverse bias, and to the junction temperature. (See Fig. 1-7). Snubber circuits can modify the load line in a favorable manner.

From a practical standpoint, a simple RC (resistive-capacitive) snubber connected across the collector-emitter terminals of the switching transistor can be quite effective in absorbing the energy of switching transients, which might otherwise damage the transistor via reverse-bias secondary breakdown. For many applications, a 300 ohms, 0.02 μF (microfarad) combination is a good starting point. The resistance should be noninductive and be capable of dissipating 10 W. From this combination, it is usually possible to optimize final RC values. The procedure

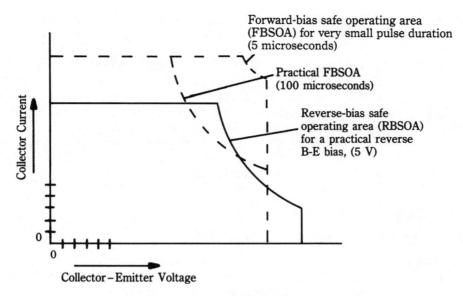

1-6   Safe operating areas for bipolar power transistors.

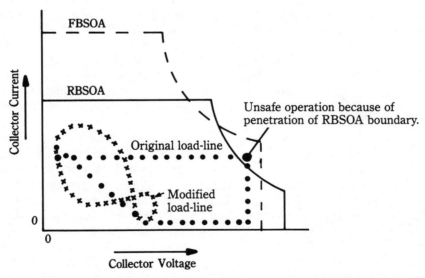

1-7   Load line of bipolar transistors must not penetrate SOA boundaries.

is to monitor the collector-emitter switching wave, paying heed to both fall time and the switching transient following turn off. Because of circuit strays, the switching transient tends to be oscillatory, even if the load is essentially resistive. An oscilloscope with good high-frequency response might be necessary to see the true nature of this transient. The objective is to find an RC combination in the snubber that exerts the greatest damping of the switching transient, while minimally affecting the turn-off time of the voltage wave.

The energy dissipated in the resistive element of the RC snubber must be supplied by the switching transistor during its forward-biased interval and will lower the operating efficiency of the switching circuit. These, however, are considered worthwhile trade offs; the transistor is more electrically rugged while forward biased, and the degradation in efficiency need not be excessive if the circuit is not over snubbed. In any event, a skimpy heat sink would be counterproductive in efforts to operate the transistor within its SOA boundaries. Also, care should be taken that transients on the ac utility line do not appear on the dc supply; these, sometimes, can be the straw that breaks the camel's back.

Transient protection for bipolar switching transistors

Figure 1-8 shows commonly encountered protective circuits for bipolar power transistors used in switching applications. Often, the terms snubber, damper, and clamp are used interchangeably. Although dampers and clamps limit load-line excursions, snubbers exert more modification of the entire load line. Figures 1-8D and 1-8E are examples of snubbing networks. Figures 1-8A, 1-8B, 1-8C, and 1-8G are examples of clamping techniques, and protective circuit I-8F is usually called a damper. This circuit is much used in the horizontal output stage of TV sets.

All of these protective circuits share the common feature that, when properly deployed, destruction by secondary breakdown can be avoided. Such destruction otherwise tends to occur in switching applications when the energy stored in load or circuit inductance penetrates the RBSOA boundary of the transistor when it is turned off. Protection is accomplished via absorption of much of this excess energy.

Figures 1-8A and 1-8B are nearly always used with inductive loads such as motors and solenoids. The inclusion of series resistance R_s can provide additional dissipation of excess energy when the transistor is turned off. Over a wide range, the amount of inductance in the switching circuit is not the governing factor of RBSOA vulnerability to destruction; rather, it is the time constant of the inductive circuit, given by L/R_s. Note that the insertion of R_s reduces this time constant. R_s can also be viewed as a Q spoiler of the resonant circuit set up between L and circuit stray capacitance. High Q resonant circuits are reservoirs of high values of circulating energy; it is, of course, preferable to dissipate LC (inductive-capacitive) energy in a resistance than in the collector junction of the transistor. In many practical circuits, there is often sufficient resistance in the inductive load itself to serve this function. That is why circuit A of Fig. 1-8 is seen more often than circuit B.

In circuit C, an appropriately selected zener diode protects the transistor by clamping the collector-emitter voltage at, or below $V_{CEX(sus)}$. Although all zener diodes are fast acting, it is preferable to use zener diodes specially made for absorption of transient energy, such as Motorola's line of Mosorbs. Circuit G is commonly encountered in automotive ignition systems. Here, FBSOA is worsened

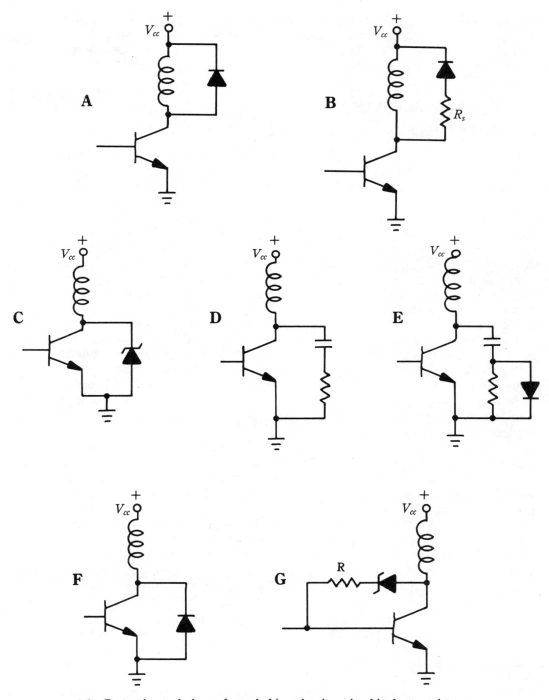

1-8 Protection techniques for switching circuits using bipolar transistors.

because RBSOA energy regeneratively extends the on time of the transistor. That is considered a worthwhile trade off because transistors tend to be more electrically rugged in their FBSOA modes than when forced to absorb RBSOA energy. Some of the excess RBSOA energy is dissipated as heat in resistance R.

Switching transients in power MOSFET circuits

When power MOSFETs are used in switching applications, a particularly advantageous feature is that they, unlike bipolar transistors, are not susceptible to secondary breakdown destruction. This feature is illustrated in Fig. 1-9. It can be readily seen that the load-line limits imposed by the shaded area do not exist for the power MOSFET. This is especially rewarding in the high-voltage, low-current operational region where the bipolar transistor exhibits high vulnerability to secondary breakdown. Inasmuch as neither FBSOA or RBSOA boundaries are stipulated for the power MOSFET, it is often considered the more rugged and more forgiving switching device. However, certain precautions still must be observed, and it is common to incorporate similar protective circuits to those described for use in bipolar transistor switching applications.

At the voltage boundary indicated in Fig. 1-9, an avalanche condition takes place. This voltage boundary, $V_{(BR)DSS}$ is somewhat analogous to the bipolar transistor boundary, $V_{CEX(sus)}$. However, it is the better part of wisdom to avoid drain voltage avalanche in the power MOSFET. That is why snubber and clamping techniques are often used. Even though it is easier to operate power MOSFETs safely without such protective networks than bipolar transistors, protection should at least be used during the experimental and breadboarding phases of circuit development.

A significant point of difference between MOS and bipolar devices is the vulnerability of the former to destruction of the gate by static charge during handling, or by voltage transients during operation. Fortunately, the relatively large gate-source capacitance makes the power MOSFET less vulnerable to destruction from electrostatic discharge than signal-power MOS devices. During experiment at least, it is a good idea to connect a 20 V zener diode between the gate and source terminals to protect against voltage transients. Such transients may come from the power supply, or may be internally transferred from the drain circuit. Once a power MOSFET switching circuit has been placed in proper operation, gate damage is unlikely to occur. Although very thin, the silicon dioxide gate insulation compares favorably with the highest quality capacitor dielectric materials.

Further examples of protection by clamp diodes

Figure 1-10 shows commonly encountered switching circuits with accompanying clamping diodes to prevent inductive kickbacks from damaging the power switch-

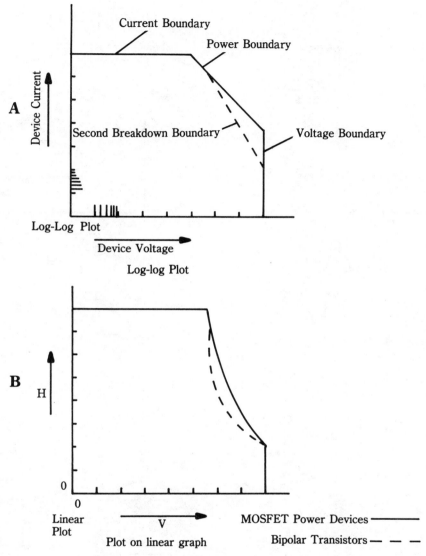

1-9 General difference in SOA between MOSFETs and bipolar transistors.

ing devices. Keep in mind that the inductance can be associated with the load even though the load might be conventionally depicted as a resistance in the schematic in diagram. Also, a motor or a transformer can essentially appear resistive, but might involve an appreciative inductive component. Even a well-designed transformer feeding into a resistive load will present sufficient leakage inductance to produce a potentially dangerous inductive kick when driven by square waves or by PWM (pulse-width modulation) waveforms. And, some circuits contain filter

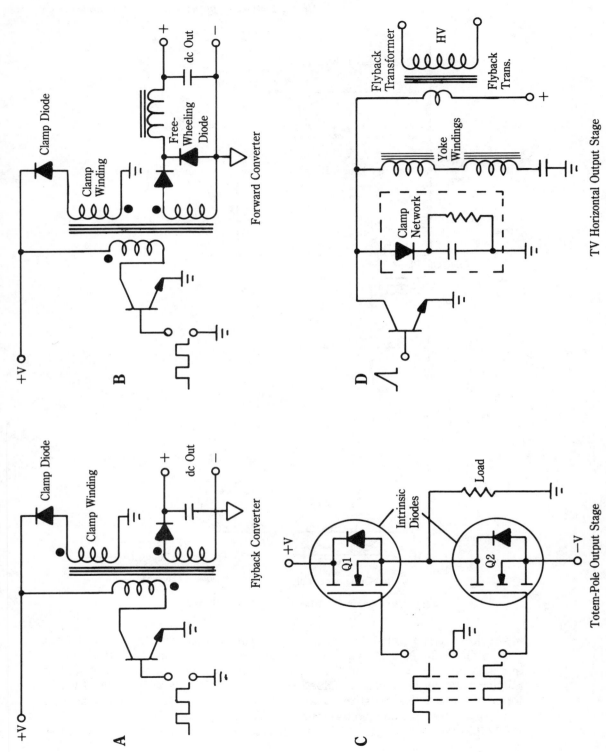

A Flyback Converter

Clamp Diode

Clamp Winding

dc Out

+V

B Forward Converter

Clamp Diode

Clamp Winding

Free-Wheeling Diode

dc Out

+V

C Totem-Pole Output Stage

Intrinsic Diodes

Q1

Q2

Load

+V

−V

D TV Horizontal Output Stage

Flyback Transformer

HV

Flyback Trans.

Yoke Windings

Clamp Network

1-10 Transient protection via clamping diodes in common switching circuits.

chokes that can cause the switching transistor to see an LR (inductive-resistive), rather than a resistive load. Finally, parasitic inductance in the PC (printed-circuit) board, itself, can give rise to switching transients of sufficient magnitude to cause trouble at high switching rates.

In Fig. 1-10A, the flyback converter makes use of a clamping winding and a clamp diode to protect the switching transistor. A similar protective arrangement is used in the forward converter circuit of Fig. 1-10B. Note the transformer phase dots in the two circuits and the difference in operating mode as indicated by the waveforms. Moreover, the filter choke and free-wheeling diode of the forward converter are *not* required in the flyback converter. In both circuits, two separate grounding systems enable advantage to be taken of the isolation between input and output provided by the transformer. Of course, if such isolation is not mandatory, a single-grounding system can be used throughout. In many instances, the clamping winding has the same number of turns as the primary winding, and these two windings are wound in *bifilar* (using two wires) fashion to minimize leakage inductance.

The clamp winding and clamp diode of the forward converter of Fig. 1-10B also serve to reset the core, a needed function in this type of switching circuit. Because it takes the same time to set and reset the core of the transformer, the duty cycle of this circuit cannot exceed 50 percent. The basic difference between the lookalike circuits of the flyback and forward converters is that the flyback converter delivers energy to the output winding when the transistor turns off; conversely, the forward converter delivers energy to the transformer output winding during on-time of the switching transistor.

The totem-pole output stage of Fig. 1-10C uses the intrinsic diodes of the power MOSFET structure for clamping inductive kickbacks. These diodes are an integral part of the power MOSFET whether or not the drafter depicts them in symbols. Depending upon the circuit, the intrinsic diode can be passive, useful, or can prevent proper operation. In the totem-pole circuit shown, the intrinsic diode of Q1 clamps the kickback produced when Q2 turns off; similarly, the intrinsic diode of Q2 clamps the kickback produced when Q1 turns off. The nice thing about this arrangement is that no external clamp diodes are needed.

The simplified circuit of a TV horizontal output stage shown in Fig. 1-10D contains a clamp network essentially equivalent to that of Fig. 1-8E. Such clamp networks provide a momentary low-impedance path for the inductive kickback transient without absorbing needless energy beyond this function. Such clamp networks allow greater switching efficiency than does the simpler RC transient suppressor of Fig. 1-10D. The resistance R of Fig. 1-10D only needs to discharge capacitor C between pulses. Complete discharge is generally unnecessary, and the value of R tends to be noncritical. It is often in the several kilohm range, as opposed to the resistance in Fig. 1-8D, which commonly ranges from several tens to several hundreds of ohms. R in Fig. 1-10D need not be noninductive in most cases.

Snubber network values

The RC snubber of Fig. 1-8D can be empirically designed by starting with several hundred ohms of noninductive resistance, then determining the minimum capacitance that satisfactorily minimizes the turnoff transient; then further experimentation with R and C can optimize results. Although empirical procedure is also rewarding for Fig. 1-8E, it is best to first arrive at ballpark values by calculating R and C.

For the capacitor, this objective is attained by means of the relationship:

$$C = \frac{(I)\,(t_f)}{V}$$

where I is the peak switching current, t_f is the fall-time of the switching transistor, and V is the peak switching voltage.

The fall time, t_f, is readily available from the specifications of the device. Interestingly, algebraic rearrangement of this equation yields:

$$CV = (I)\,(t_f)$$

where the quantities on both sides of the equal sign represent Q, or charge. Translated into circuit operation, the capacitor absorbs the change in charge produced when the transistor turns off. This is tantamount to stating that the turn-off transient is transferred from the switching device to the capacitor.

Note also that V tends to be approximately twice the power-supply voltage for single-transistor switching circuits such as those depicted in Figs. 1-8D and 1-8E. (In contrast, V is equal to the power-supply voltage when push-pull, half-bridge, and full-bridge switching circuits are used.)

Now, what about R for the snubber network of Fig. 1-8E? It turns out that R is a function of the minimum on time of the switching waveform, and can be determined from:

$$R = \frac{t_{on}}{C}$$

Where t_{on} represents the minimum pulse-duration, or on time.

Also, the power rating of the resistance, R, is given by

$$P = \frac{CV^2(f)}{2}$$

In all of the foregoing relationships, R is expressed in ohms, C is expressed in Farads, V is expressed in volts, I is expressed in amperes, P is expressed in watts, f is expressed in Hertz, and t_f and t_{on} are expressed in seconds. This is the dimensional standard advocated by most textbooks on physics.

Phase-control facts

Power systems using thyristors often exert control of load power by control of the time of firing during the ac cycle. This implies delaying conduction of the thyristor so that only a selectable portion of the ac cycle is available for the load. The non-sinusoidal voltage and current wave applied to the load has operational, measurement, analytic, and harmonic interference ramifications not encountered with straight dc or with ac sine-wave power. However, such phase-control circuits are relatively simple, efficient, and very convenient. Efficiency is high because turn on is extremely fast, and turn off is reasonably fast. Also, conductive losses tend to be low because of the volt or so dropped across these devices. Manufacturers provide a wide variety of thyristors, from signal-level types to giant devices capable of handling kilovolts and kilo-amperes simultaneously. A nice thing about using thyristors for phase-controlled power is the self commutating feature, wherein turn off automatically obtains when the current wave goes through zero. Phase-controlled power can be accomplished either on a half-wave or on a full-wave basis. A single SCR (silicon-controlled rectifier) yields a half-wave circuit, whereas triacs enable full-wave operation from a single device.

Several relationships pertaining to phase-controlled power are often overlooked or become the sources of confusion. As might be suspected, a half-wave circuit can only deliver half the load power forthcoming from a full-wave circuit. However, it follows that the RMS (root mean square) current capability of a half-wave controller is then 70.7 percent, not 50 percent of a full-wave system. By the same token, the maximum RMS voltage that a half-wave circuit can deliver to the load is 70.7 percent that of a full-wave controller; the full-wave technique can deliver just about the same RMS load voltage as the RMS line voltage.

There is often needless worry that sufficient control range will be attained by a phase-control system. Such apprehension is generally unfounded for the following reasons: in a full-wave control circuit, a conduction angle of 30 degrees corresponds to only 3 percent of full-load power; a conduction angle of 150 degrees provides 97 percent of full-load power. This means that 94 percent of full power control is available from a phase adjustment of only 120 degrees. Effort to produce wider phase adjustment obviously leads to greatly diminishing returns. See Fig. 1-11 numbers for a half-wave control situation are different, but likewise lead to the conclusion that phase-control throughout the 30- to 150-degree range suffices for practical power control.

Thyristors, too, can misbehave as switches

Thyristors also have switching problems that can be improved by snubbing networks, usually of the simple RC type of Fig. 1-8D. In contrast to the switching problems of bipolar transistors and power MOSFETs, it is the extremely rapid turn on of thyristors that tends to cause troubles. For example, the RFI (radio frequency interference) and EMI (electromagnetic interference) generated by thyristors originates primarily from this characteristic. Such interference might not only play havoc with communications equipment, but all too often causes false turn on of other thyristor control circuits. Switching waveforms in phase-controlled thyristor power systems are shown in Fig. 1-12.

Another source of false turn on in thyristors emanates from the so-called *dv/dt* effect. In so many words, if the voltage across the thyristor rises too rapidly follow-

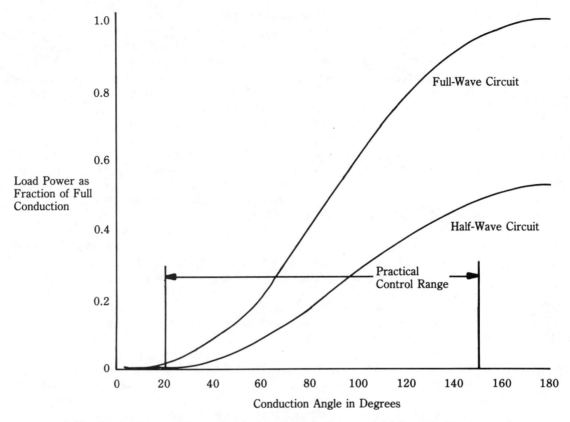

1-11   Load power versus conduction angle for half-wave and full-wave phase control.

ing turn off, the thyristor might be internally retriggered, thereby losing control. This retriggering occurs because of capacitive feedback from the anode to the gate and is one of the factors limiting the frequency at which proper control can be maintained. High frequency is tantamount to high *dv/dt* and there is more internal capacitive feedback to the gate. In actual applications, snubbing is very effective in preventing erratic performance of this nature. Of course, the proper thyristor must be selected for the frequency involved.

Another way of improving *dv/dt* immunity is to use a low value of gate-cathode resistance to divert much of the internal anode-gate feedback current. This is a very effective approach, but it is at the expense of increased drive power. A negative bias of about 1 V at the gate can also be used to extend immunity from the *dv/dt* effect. Such reverse bias is more easily applied to SCR than to triac circuits.

Whether false triggering is attributed to RFI, line transients, or *dv/dt* effects, keep in mind that thyristors trigger with less gate charge as temperature increases. A basic requirement of a satisfactory driver is to adequately trigger the thyristor over its entire range of operating temperature. This sometimes imposes a problem at the cold end of the range where it is found the thyristor will not turn on. Con-

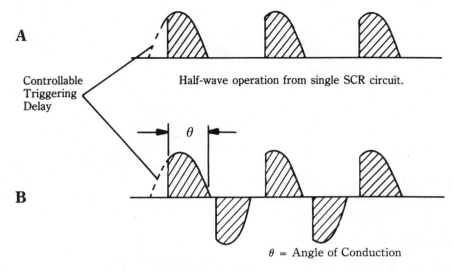

A

Controllable
Triggering
Delay

Half-wave operation from single SCR circuit.

θ

B

θ = Angle of Conduction

Full-wave operation from single-triac, or from back-to-back SCRs
(anti parallel) circuit.

1-12   Load-voltage waveforms in thyristor phase-control systems.

versely, at the high temperature region of operation, there may be unanticipated vulnerability to false turn on.

False turn on problems are made worse by inductive loads; the thyristor tries to turn off while there is high voltage across it because of the phase displacement of voltage and current caused by the inductive load. That is, zero load current no longer coincides with zero load voltage as is the case with a resistive load. This is essentially a high dv/dt situation tending to retrigger the thyristor as previously described. This type of dv/dt, known as commutating dv/dt can exist in 60 Hz and low frequency control systems, but otherwise acts as ordinary dv/dt. The solutions to this type of false triggering are, for the most part, those already mentioned—snubbing for reducing dv/dt and gate circuit modifications for reducing vulnerability to the internal anode-gate feedback current.

As if all this is not enough, thyristors can be damaged from too high a rate of load current rise during turn on. Turn on can be likened to the ignition and propagation of a flame, which is not instantaneous. Making a high current demand before the junctions are fully aflame can create localized hot spots in the junctions not unlike those causing secondary breakdown in bipolar transistors. Obviously, this type of damage is not likely to occur with inductive loads where the initial rate of rise of the current is impeded. By the same token, a guard against di/dt destruction when using resistive loads is to insert a small amount of inductance in the load circuit. Sometimes a delay reactor in the form of a nonlinear inductance is used. The square-loop core permits initial slow-down of di/dt. (See Fig. 1-13.) A straightforward way of improving the di/dt capability of a thyristor is to drive it hard from a rapid-rise trigger source. This speeds up the propagation process so that full turn-

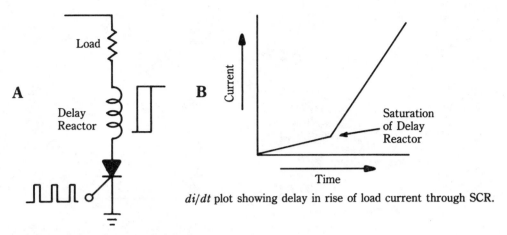

A Partial circuit of radar modulator with delay reactor.

B *di/dt* plot showing delay in rise of load current through SCR.

Load

Delay
Reactor

Saturation
of Delay
Reactor

Current

Time

1-13 Use of a delay reactor to slow initial load current.

on capability is quickly attained. Conversely, *di/dt* failures are usually found in circuits where the gate drive is minimally sufficient to reliably trigger the thyristor. Within reasonably limits, the only drawback to overdriving the gate is to slightly lower overall switching efficiency.

Dispelling an illusion about power factor

Table 1-3 depicts the operating parameters for several conduction angles in a full-wave phase-control circuit. It is commonly realized that because of the nonsinusoidal wave shapes, simple arithmetic relationships of current, voltage, and power do not prevail. Nonetheless, neophyte and old timer alike often encounter a stumbling block with regard to the power factor seen by the ac utility line. With a purely resistive load, it is true that the load power factor remains unity for all conduction angles. Because no inductive or capacitive reactance is involved, it is all too often supposed that the line power factor must also be unity over the power control range. As the table reveals, this is far from the truth of the matter.

Inasmuch as a low power factor at the ac line implies a higher line current than would exist at unity power factor, the overall efficiency of a phase-controlled power system can be significantly less than the published figures by marketing departments. For practical purposes, the switching efficiency of the thyristor itself can be multiplied by the power factor of the ac line in order to obtain the true efficiency of the system. Thus, a triac circuit capable of 90-percent efficiency at a conduction angle of 90 degrees would look like a 0.90 × 0.7, or approximately a 63-percent efficient circuit to the ac line. From the standpoint of the utility company, it makes little difference whether low power factor emanates from actual physical inductance, or from a distorted current waveform. In both cases, there will be increased voltage drop and heating in the ac supply lines. In both cases, more actual ac RMS current must be supplied than would be necessary with a unity power factor at the line.

Table 1-3. Waveforms and parameter values in full-wave, phase-control system.

Line-voltage waveform (sine wave)	Conduction angle in degrees	Line and load current waveform (load voltage waveform)	Effective current (RMS value)	Apparent power in load	True power in load	Power factor at ac line	Power factor at load
Em	180	Im	0.707 Im	0.500 EmIm	0.500 EmIm	1.000	1.000
	150		0.697 Im	0.493 EmIm	0.486 EmIm	0.985	1.000
	120		0.634 Im	0.448 EmIm	0.402 EmIm	0.897	1.000
	90		0.500 Im	0.354 EmIm	0.250 EmIm	0.707	1.000
	60		0.313 Im	0.221 EmIm	0.098 EmIm	0.443	1.000
	30		0.120 Im	0.085 EmIm	0.014 EmIm	0.170	1.000

As might be supposed, the above-described situation is even worse for half-wave phase-controlled circuits using single SCRs. For both, full-wave and half-wave control systems, the ac line power factor is given by:

$$\frac{\text{True power}}{\text{Apparent power}}$$

The measurement problem involves the use of instruments capable of indicating true power, and RMS voltage and RMS current, the product of which is apparent power. Many of the popular texts on electrical engineering gloss over subject of power factor resulting from distorted wave shapes of voltage or current. It is interesting, and relevant to thyristor circuits, to note that unity power factor exists in a resistive load if both load current and load voltage have the same wave shape.

An elusive transformer problem in PWM systems

Pulse width modulation is a popular way of controlling power when working from a dc source. This method is widely used in regulated power supplies, in electric vehicles, and in other motor-control systems. Load power control is accomplished by varying the on or off time of a switching device, then averaging or filtering the chopped waveform. High operating efficiency can be realized for the simple reason that the switch is either in its on or in its off state of conduction, with minimal time allotted to the transition between the two conduction states. A peculiar trouble often arises because of a common oversight in the driver transformer for such a PWM switch.

These days, relatively few engineers have had extensive exposure to the theory and design of transformers because training focus has been on computer themes. Moreover, even those who learned about transformers used texts that generally confined discussion to sine waves, and sometimes square waves when electronic circuits were studied. The insights presented by such constraints are fine but lead to invalid assumptions when the duty cycle of the square wave is other than 50 percent. Thus, in a PWM waveform, where the duty cycle is constantly varying, transformer operation departs from both sine-wave and symmetrical square-wave behavior. This is where the mysterious poor performance of PWM systems often asserts itself, and the reason can be quite elusive if you do not know what to look for.

Figure 1-14A shows sine- and symmetrical square-wave operation of a transformer. Assuming a one-to-one ratio between the windings for simplicity, the voltage waveform across the secondary is, essentially, a replica of the waveform impressed across the primary winding. For a square wave, a certain amount of distortion can be visible at low frequencies, but the symmetry of the secondary waveform remains intact. The dc component of the primary waveform, if any, is lost in the secondary because a transformer cannot transfer dc. This is usually understood and need cause no trouble.

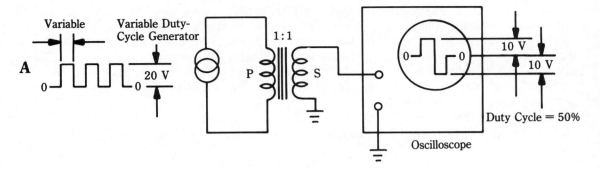

A

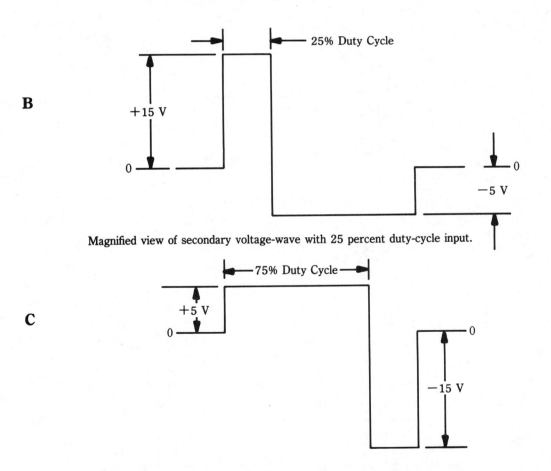

Demonstration setup operating with 50 percent duty-cycle input wave.

B

Magnified view of secondary voltage-wave with 25 percent duty-cycle input.

C

Magnified view of secondary voltage-wave with 75 percent duty-cycle input.

1-14 Duty-cycle effect on the secondary voltage of a transformer.

The situations depicted in Fig. 1-14A and B are another story; when the duty cycle at the primary is no longer 50 percent, the amplitudes of the positive and negative excursions of the secondary voltage are no longer equal. If, as is often the case, a power MOSFET is being driven from the secondary winding, the enhancement voltage for the gate will vary with the duty cycle. Thus, the power MOSFET is likely to operate from hard-driven to under-driven conditions. In the latter case, its drain-source voltage drop will increase, giving rise to faulty operation and excessive power dissipation. Unfortunately, if diagnosis is made under the wrong duty-cycle drive conditions, the reason for the poor performance might remain concealed.

If a wide range of control is required from a PWM format, something must be done to keep the amplitude shifts of the secondary voltage in the drive transformer within bounds. Otherwise a driven power MOSFET can either be destroyed by breakdown of its gate insulation, or can be subjected to overheating because of insufficient gate drive. A brute-force remedy is shown in Fig. 1-15A. Here, the

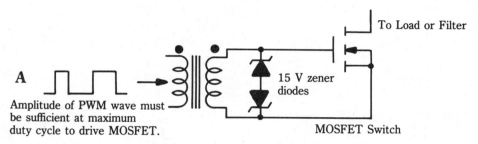

Zener diodes in brute-force clamping circuit.

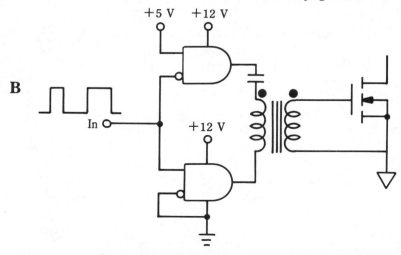

Circuit using logic gates impresses true bilateral waveform on transformer primary regardless of duty cycle.

1-15 Remedial circuits for transformer distortion of PWM waveform.

positive and negative excursions of the secondary voltage in the driver transformer are both clamped at approximately 16 V by zener diodes. (15 V of zener breakdown voltage plus 1 V due to forward voltage drop in the alternate zener diode results in an effective clamp of about 16 V.) Most power MOSFETs saturate in the vicinity of 10 V applied to the gate but tend to develop a lower R_d at a somewhat higher gate voltage. In any event, it is usually necessary to keep the gate-source voltage well below twenty volts in either polarity.

An objection to the technique of Fig. 1-15A is that it dissipates drive power, making greater demand on the transformer size and the driver circuit. After all, one important reason for using the power MOSFET as a switch is that relatively little drive power is needed to put it through its paces. With this in mind, a more sophisticated approach is shown in Fig. 1-15B. Inspection reveals that something is done to the primary, rather than the secondary, circuit of the transformer in order to make the positive and negative induced voltage independent of duty cycle. This is realized through the operation of a logic circuit that presents the primary of the transformer with a symmetrical bipolar waveform while being actuated from a unipolar PWM source.

Note that when one gate is on, the other is off, and there is no possibility of overlap of their conduction states. Suitable logic gates can be used from the D469 IC, which actually comprises four independent gating circuits. An additional secondary winding on the transformer can be used to drive a second power MOSFET of a totem-pole output stage. A negative power supply would then also be needed. This scheme can provide trouble-free operation over a wide duty-cycle range because the transformer always sees a true ac wave.

Integral cycle control of load power

Pulse-width modulation and phase control are not the only way control of power can be achieved via solid-state switching devices. Thyristors can efficiently control load power by the integral-cycle technique, also known as burst modulation. In this scheme, the thyristor is switched to its fully on state (180-degree angle of conduction), then turned off again for a number of integral cycles of the applied ac line voltage. (See Fig. 1-16.) Either, or both on and off periods can be manually adjusted or automatically varied by a sensing and feedback circuit. Heaters, because of their large thermal inertia, are particularly suitable for this control technique. Motors, too, can be satisfactorily controlled with a little extra care in design parameters. In such systems, the motor is not subjected to nonsinusoidal waveforms of the applied voltage. This reduces both electrical and mechanical losses and results in a quieter-running motor with less stress in its bearings.

Integral-cycle control requires triggering at zero line voltage and the process is, appropriately enough, often referred to as zero-voltage switching. It is possible to use discrete components in circuits for triggering thyristors at zero line voltage, but it is generally more convenient to use one of a number of available ICs designed for this purpose. The scheme, in any event, is best suited for use with triacs because of their full-wave operation. The salient features of integral-cycle

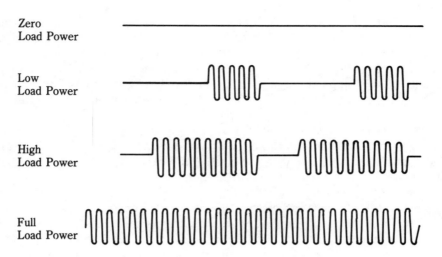

These are idealized waveforms for resistive loads, such as heaters. Satisfactory control of motors is possible if time-constant L/R is low.

1-16 Voltage delivered to load by a triac in an integral-cycle power controller.

control of load power are as follows:

1. There is less stress on the thyristor because of greatly reduced dv/dt, and for some loads, di/dt as well.
2. There is significantly less RFI and EMI. Often the RFI filter can be omitted. Altogether, interference from shock excitation and from harmonic energy are virtually eliminated.
3. False triggering from commutation dv/dt is eliminated.
4. Snubber requirements are relaxed; often snubber circuits can be omitted.
5. Certain loads operate under less abusive conditions. The initial inrush current of incandescent lamps and of motors is greatly reduced, thereby prolonging life.
6. Temperature effects on triggering are reduced, facilitating design and operation.
7. Measurements can be made with ordinary instruments.
8. Unity power-factor operation with resistive load is possible.

The more nearly resistive the load, the easier is the implementation of this switching technique. Motors often require extra considerations, but they, too, becomes essentially resistive loads when loaded.

It is only natural that a circuit technique possessing such features as integral-cycle control, or zero-voltage switching, must have some drawbacks, as well. Indeed, one disadvantage of this power-control method is its behavior with inductive loads. The worst time to introduce ac power to an inductive load is during the zero-cross interval of the applied voltage. Such timing produces the greatest surge current. The transient so produced in essentially a dc component and a time-decaying asymmetry of the ac current. (See Fig. 1-17.) Motors usually manifest themselves as equivalent LR circuits; it turns out that control via zero-voltage switching can be practical if the L/R time constant is not too large. This implies that a partially loaded motor can cause less trouble with initial current surges than an unloaded one. (This reasoning is more valid for small than for large motors because starting current for large motors can be very high even without zero-voltage switching.)

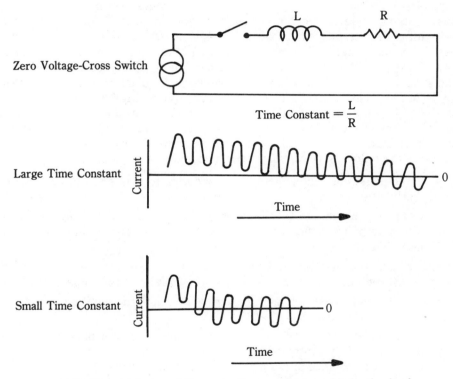

1-17　Transient created by zero-voltage switching of inductive load.

Another disadvantage of integral-cycle control is that benefits otherwise associated with the frequency of the chopped carrier are not forthcoming, but are limited by the average rate of interruption of the wave trains. Thus, a regulated power supply providing power made up of interrupted 20 kHz wave trains would be much harder to filter than conventional 20 kHz supplies; if the average interruption rate were in the vicinity of 1 kHz, the electrical and physical parameters of filter components would be dictated by 1 kHz ripple. Also, such a supply would show only the dynamic response of a 1 kHz switching rate.

A somewhat controversial topic concerns the matter of implementing a stable feedback network for regulation purposes. The on and off nature of integral-cycle control can indeed introduce difficulties not encountered in systems operating at fairly steady levels. However, much depends upon the technical and experimental skills of the designer. The easiest integral-cycle systems to handle are probably those associated with heaters; these systems involve resistive loads, low frequencies, and low interruption rates. Furthermore, they are generally not high-precision systems, requiring only moderate feedback percentages.

Polyphase versus single-phase power

Polyphase power systems have a number of advantages over more simple single-phase systems. Indeed, it is not always true that simplicity is necessarily associated with the single-phase approach. The advantages are well known because of the long exploitation of the three-phase format in utility and industrial power systems. Because of shortcomings in power-handling capability, reliability, and cost, solid-state power techniques were slow to make their appearance in polyphase systems. That era is over. Bipolar power transistors, power MOSFETs, and thyristors have evolved to the point where respectable power levels can be implemented. At the same time, logic devices and techniques make waveform synthesis feasible with surprisingly few IC modules. All in all, polyphase power systems can be said to possess the following advantages over the single-phase format:

- Three-phase induction motors are superior to single-phase ac motors in nearly all respects.

- Two-phase motors are excellent for use in servo systems. They are cheaper and more maintenance free than are dc motors. And, unlike single-phase induction motors, they are self-starting.

- For airborne and space applications, three-phase transformers provide weight reduction over single-phase units working at the same power level.

- Rectified polyphase voltage is easier to filter than the rectified voltage from a single-phase source.

- A three-phase inverter or power supply maintains power balance in a three-phase source. By contrast, single-phase inverters or power supplies contribute to the imbalance of power in a three-phase power source.

- Three-phase transformer windings can be connected so that much of the third-harmonic energy in a square or quasi-square wave is attenuated. Such electrical or electronic filtering reduces the need to rely upon bulky physical filters. This is of considerable importance in the operation of motors; harmonics can cause temperature rise and rough operation. The cancellation of the third harmonic also removes higher-order odd harmonics that often cause electrical interference with other systems.

- The manner of connecting the windings of three-phase transformers also provides the designer and user with additional flexibility in the selection of output voltage and available current.

- Logic circuitry for generating the polyphase format can be readily manipulated to provide various performance features. These include dead-time intervals, stepped shapes to simulate sine waves, changing phase sequence for reversing induction motors, harmonic manipulation, frequency variation for speed control of induction and synchronous motors, and output-voltage control. Such logic circuitry also facilitates feedback applications for maintaining constant speed, torque, or horsepower.

Relationships in three-phase power systems

The basic source or load connections encountered in three-phase systems are shown in Fig. 1-18. In these illustrations, resistive loads are shown connected in symmetrical formats to receive balanced power from their three-phase sources. It is assumed that the three loads in each diagram are equal. Because of the phase relationships of the three source voltages, common-sense arithmetic leads to erroneous calculations. For example, in the delta connection of Fig. 1-18A the line current, *I*, is not three times the current that flows in each load element. Rather, the line current is 1.732 times an individual load current. This is a basic relationship and derives from the trigonometric relationships in three-phase systems. Similarly, in the Y connection of Fig. 1-18B, phase voltage or line voltage is 1.732 times the voltage across an individual load. The number 1.732 is an ever-present factor in

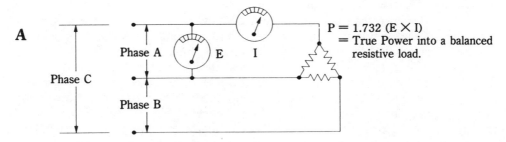

Delta-connected loads receiving three-phase power.

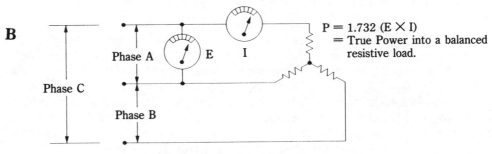

Y-connected loads receiving three-phase power.

1-18 The basic connections used in three-phase circuits.

three-phase calculations; it is the square root of 3 and is also the tangent function of 120 degrees.

Note that the phase voltages in the delta connection are equal to the individual load voltages. Here is a case where you can state that the equality is obvious from inspection and not have to suffer subsequent embarrassment for ignoring some subtlety of mathematics. Reciprocally, you can see that the line current and the individual load currents in the Y connection are the same.

From the above considerations, it turns out that total load power, P, in both delta and Y systems is given by $P = 1.732 (E \times I)$, with E and I as depicted in Fig. 1-18. The common mistake made by those not familiar with three-phase relationships is to suppose that $P = 3 (E \times I)$, a natural enough error, deriving from experience with single-phase systems.

If the loads have reactance as well as resistance, the equation for true power, P, becomes $P = 1.732 (E \times I)(PF)$ where PF is the power factor of the load impedances. One definition of power factor is that it is the ratio obtained by dividing true power by apparent power. Apparent power, when the loads are impedances, is the previous expression, $P = 1.732 (E \times I)$. True power can be obtained from wattmeter measurements. However, if the impedance, Z, and resistance, R, of the load elements are known, the wattmeter measurements are not needed. This is because $PF = R/Z$. Summarizing, true power = $1.732 (E \times I)(R/Z)$ for all load impedances in both delta and Y systems, where Z is the individual load impedance, and R is the individual load resistance.

Polyphase power can be likened to a multicylinder engine—there is a more continuous flow of power than in a one-cylinder (single-phase) energy source. Although this analogy might not be very meaningful for resistive loads, a polyphase motor does have a smoother rotational torque than a single-phase motor. This has been one of the salient features of three-phase motors in industry. Now, via power electronics, such motors can be smoothly controlled with relatively inexpensive equipment. This is particularly significant for electric vehicles and traction applications, for it has become feasible to convert either dc or single-phase power into the polyphase format (usually three phase) by solid-state power devices.

Format of the three-phase power system

Considerable insight into the nature and behavior of three-phase systems can be gained by a close scrutiny of the three 120-degree displaced sine waves shown in Fig. 1-19A. At any instant of time, the algebraic sum of the three waves is zero. This is important and is not obvious from casual inspection. A near proof of this relationship is as follows. Consider the situation when one of the waves is at its maximum value. At that time, it will be seen that the other two waves are both at 50 percent of their maximum values, but of the opposite polarity with respect to the first wave. Thus, the algebraic sum of the three waves at that time is zero. Again, consider the situation when one of the waves is of zero amplitude. At that time, it will be seen that the other two waves are at approximately 86 percent of their respective amplitudes, but of opposite polarity to each other. As before, the

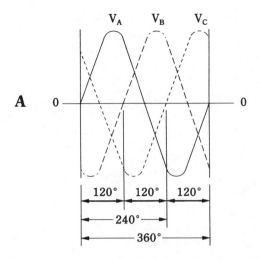

Oscilloscope display of three-phase voltage.

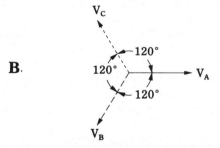

Vector diagram representation of the sine waves of Fig. 1-19A.

1-19 Format of voltages in a three-phase power system.

algebraic sum of the three waves is zero. This relationship holds for any time between 0 and 360 degrees for any single wave. More simply stated, the relationship is true at any time. The vector representation shown in Fig. 1-19B provides additional insight into the three-phase format of sinusoidal voltages.

A practical consequence of this relationship is shown in Fig. 1-20A where the coils represent either the output windings of an alternator or the secondary windings of transformers. The coils, in other words, are sources of three-phase voltages. The closed configuration, known as the delta connection, superficially appears to be a short circuit. But if the delta were opened, as depicted in Fig. 1-20B, zero voltage would exist between terminals X and Y. This is in accordance with the logic developed in the preceding paragraph. Thus, it is possible to close the delta. But two things must be kept in mind.

First, the delta can only be closed for sine waves, or for other waves having no third, sixth, or ninth harmonics (or higher integral orders of third-harmonic

A

Delta connection for three 120-degree displaced voltages.

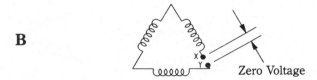

B

Zero Voltage

In a proper delta system, no voltage exists between any of the coils at their
corner connections.

1-20 Alternator output windings, or transformer secondaries, connected in delta.

energy). Such harmonics do not sum up to zero as does the fundamental frequency.
Therefore, the delta connection will constitute a short circuit to such harmonics.
(It is shown later how this can be advantageously used in certain situations.)

Second, three sources of 120-degree displaced voltages cannot be indiscrimi-
nately connected in the delta configuration. Attention must be given to the phasing
of the sources. In practice, this means that two of the sources can be connected
together without regard to which terminals are selected, but the remaining source
can only be connected one way. If it is connected the wrong way, a short circuit will
result.

It is well to be mindful that many electrical engineering texts deal primarily
with sine wave polyphase systems. In contrast, electronic inverters generally make
use of square, quasi-square, or stepped waveforms. What is valid practice for sine
waves might or might not be valid for these nonsinusoidal waves.

Three-phase format from an analog oscillator

A three-phase sine-wave format can be produced by cascaded amplifying devices
in a symmetrical circuit configuration such as shown in Fig. 1-21. In essence, this
amounts to a three-stage RC coupled amplifier in which the output is fed back to
the input. Inasmuch as a 360-degree phase shift is needed to produce oscillation, it
is apparent that such a scheme would not be oscillatory if only the phase shift
occurring in the op amps was effective. What is needed is an additional 180
degrees of phase shift. This is achieved via the RC coupling networks at the output
of each stage. The overall circuit oscillates at that frequency at which the total RC
phase shift is 180 degrees. This means that each RC network provides 60 degrees
of phase shift. When you consider the total per-stage phase shift—that provided by
an op amp in conjunction with that occurring in its output RC network—it turns

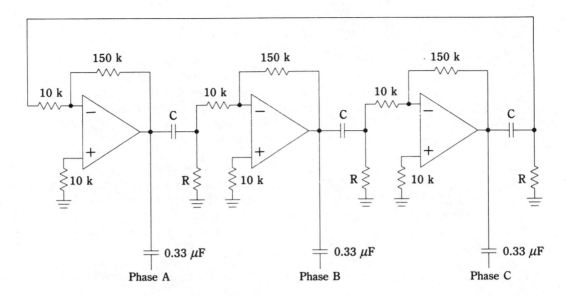

Note: Select RC to produce 120° between the phases.

1-21   Analog technique for producing a three-phase format.

out that 120 degrees of phase displacement exists between any two of the outputs. Thus, phase A is displaced 120 degrees from phase B in one direction and 120 degrees from phase C in the opposite direction. This amounts to a three-phase format of sine waves.

Note that as long as all Rs are equal and all Cs are equal, the three-phase format is preserved; only the frequency changes with a change in R or in C. Practical values of C will often fall between 0.1 and 1.0 μF over the 50 to 1000 Hz frequency range. Then, $R = X_c/\mathrm{Tan}\ 60°$, where $X_c = 1/2\pi fC$, and the tangent function of 60 degrees is 1.732. By utilizing three ganged pots, a wide control range of three-phase frequencies can be had. This is very useful for speed control of three-phase induction or synchronous motors.

The op amps can be any of the numerous designs based upon the original 741. Later versions of this op amp have FET (field-effect transistors) inputs, and there are often other sophistications. The circuit is not demanding and actually many garden-variety op amps will prove satisfactory. In any event, the basic idea is to generate good sine waves. This might require lower values than the 150 ohms feedback resistances shown. If square waves are needed, they are best obtained by signal processing, following the basic circuit of Fig. 1-21. In any event, buffer and power amplifiers are required in each phase; square waves can be produced by allowing such amplifiers to be overdriven or saturated, or by means of triggered logic devices. Whatever amplifying or wave-shaping circuits are used, they should be identical for the three phases. Otherwise, the interphase balance is likely to be destroyed, with deleterious effects on both the inverter and the load.

Three-phase format
from an RC bridge network

An RC bridge in conjunction with a center-tapped transformer can be utilized to produce a three-phase format of voltages if driven from a sine wave source. Such an arrangement is shown in Fig. 1-22. Because of tolerances in components and the difficulty of obtaining exact mathematically calculated values, some experimentation may be in order to produce three equal-amplitude voltages, spaced 120 degrees in phase. The component values depicted in the diagram will come close, however. This circuit is frequency sensitive and will not work with square waves because the harmonic frequencies will not be accorded proper phase shift or amplitude response. If square waves are needed from the three-phase format, appropriate squaring circuits can be introduced following the three-phase conversion.

Note in Fig. 1-22 that one of the four RC junctions in the bridge is not used for providing an output voltage in the three-phase format. It is important that the unused junction be the one indicated; although the schematic diagram appears to be topographically symmetrical, only the selected three junctions will provide the required 120-degree phase displacements.

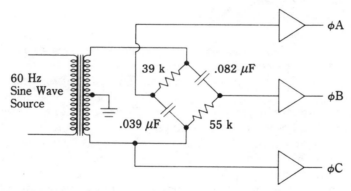

1-22   Single-phase to three-phase conversion with an RC bridge.

In the event a sine-wave frequency other than 60 Hz is involved, the capacitors in the bridge should be changed to offer the same reactance as the indicated ones do at 60 Hz. Thus, if the input frequency were 600 Hz, the capacitors would be changed to one-tenth of their 60 Hz values. The resistances in the bridge can be retained at their 60 Hz values for any frequency.

Minor changes in the amplitude of the phases can be made by varying the gain of the amplifiers. Reversed phase rotation can be accomplished by transposing any two amplifier-input connections to the bridge. Such phase rotation changes the direction of rotation of a three-phase induction or synchronous motor. It is also an important consideration when two or more three-phase sources are to be operated in parallel for increased power capability—all such sources must have the same sequence of phase rotation and this must be determined before the sources are paralleled. Otherwise, destructive short-circuit currents will flow. (The other requisites for paralleling three-phase sources of power is that all voltages should be balanced and equal, and all frequencies should be the same.)

Digitally generated three-phase format

A generally more useful way to generate the three-phase format than by analog circuits is the utilization of digital logic. Not only is the square-wave output thereby produced usually desired for inverters, converters, and power supplies, but the precision of timed waves is not dependent upon passive component values. Most digital designs make use of a three flip-flop shift register arranged as a *twisted-ring* counter. This basic scheme is shown in Fig. 1-23. Here the flip-flops are JK binaries. When this ring counter is clocked at six times the desired frequency of the three-phase format, a set of square waves is generated that complies with the timing sequence of the three-phase format.

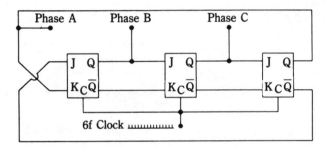

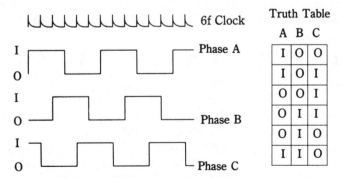

1-23 Basic principle of the ring-counter, three-phase generator.

A	B	C
I	O	O
I	O	I
O	O	I
O	I	I
O	I	O
I	I	O

The simple setup of Fig. 1-23 has a shortcoming, however. Although the truth table is essentially correct, it does not tell the whole story. In addition to the six indicated logic-level combinations, there is also the possibility of the counter locking up in either 000 or 111. Such lockup tends to occur when the counter is switched on (when dc operating power is first applied), or when a noise transient appears during ordinary operation. What is needed is additional logic to prevent the occurrence of these undesired states.

A more practical form of the digital three-phase generator is shown in Fig. 1-24. The additional NOR and OR gates prevent the counter from deviating from the truth table of Fig. 1-23. Otherwise, the performance is essentially the same as in the basic setup. Although there is a large variety of these three-phase generators using various types of flip-flops and a variety of logic gates for inhibiting the undesired counter state, most operate very similarly to the circuit of Fig. 1-24.

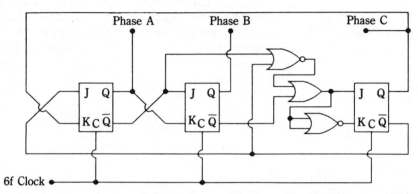

1-24 Basic circuit of Fig. 1-23 with logic gates to prevent undesired modes.

If a crystal-controlled clock is used, excellent precision in both frequency and phase balance is possible. For some purposes, the clock can be synchronized to the 60 Hz utility source. The digital method of three-phase generation merits consideration even if the ultimate output must be sinusoidal, for such conversion can be achieved via a simple LC low-pass filter, or by an active bandpass filter.

Digitally generated quadrature format

Two-phase induction motors are very useful in ac positioning and servo systems. Such motors are mechanically rugged and need little maintenance. Much like three-phase motors, they are inherently self-starting, and electrically reversible. The required voltage format for these motors is two equal-amplitude waves with the phase displacement between them equal to 90 degrees. At first consideration, such a format might appear to be unsymmetrical. Nonetheless, such a quadrature-phase format results in very smooth and efficient motor operation.

The traditional way of producing the 90-degree phase difference between the two motor windings is to insert a capacitor in series with one winding. See Fig. 1-25. Although satisfactory for many applications, exact quadrature phase displacement is not attained, and the electrolytic capacitors often used tend to change their capacity and increase their leakage current and effective series resistance with temperature and age. Moreover, if the speed of the motor is to be varied by

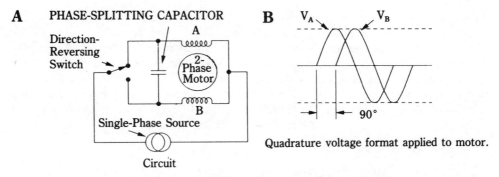

A PHASE-SPLITTING CAPACITOR

Direction-Reversing Switch

Single-Phase Source

Circuit

B V_A V_B

90°

Quadrature voltage format applied to motor.

1-25   Conventional way of producing a two-phase format for a motor.

changing the applied frequency, different capacitors must be switched in and out. This is both awkward and costly. The motor itself will usually operate well over a wide speed range providing the quadrature relationship is maintained in the two-phase supply. (For best results, it might also be necessary to increase voltage with frequency in order to maintain a constant motor current.)

An elegant way to generate a precise two-phase format for the motor is with a digital circuit such as is shown in Fig. 1-26. The two D binaries produce the accom-

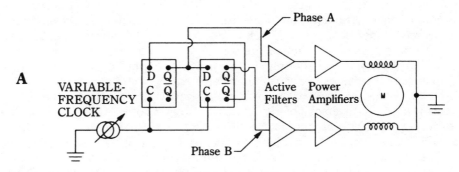

A VARIABLE-FREQUENCY CLOCK

Phase A

Active Filters   Power Amplifiers

Phase B

Circuit of digital generator.

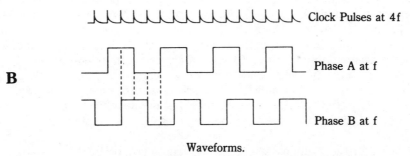

B

Clock Pulses at 4f

Phase A at f

Phase B at f

Waveforms.

1-26   Digital-logic generator for a two-phase motor.

panying waveforms of voltage. Note the requirement for 4f clock pulses. The active filters, either low-pass or bandpass, convert the square waves to sine waves. Depending upon the motor used, these filters might not be necessary. Often, the motor inductance suffices to make motor current a close-enough approximation to a sine wave so that there is no undue temperature rise or torque disturbance from the residual harmonic energy. If the filters are used and a wide frequency range is employed for speed control, it is much easier to change the RC networks in such active filters than it is to change the large capacitor in the conventional phase-shifting technique.

A diode is a diode—but not necessarily

Diodes are often the nemesis of otherwise competent designers of power-control circuits. Based upon 60 Hz experience, it is easy enough to surmise that diode selection is merely a matter of satisfying voltage, current, and power dissipation numbers. Thereafter the basic idea tends to be to install and connect these simple components in the circuit and forget about them. It often happens, however, that such safely rated diodes become the sources of many kinds of circuit poor performance and even cause catastrophic destruction of other devices and components. The tragic part of this is that the diodes themselves are not always readily pin-pointed as the culprits.

It turns out that diodes that display near-ideal characteristics at 60 Hz and at low audio frequencies might no longer perform as essentially perfect rectifiers at still higher frequencies, say at 25 kHz. This is graphically borne out in the waveform diagrams of Fig. 1-27. You are dealing here with the reverse-recovery characteristic of PN junction diodes. In terms of the physics of these diodes, instantaneous reverse recovery cannot occur because of storage of minority carriers—finite time is required for these charge carriers to recombine or to be removed when the ac cycle reverses polarity. At 60 Hz, this recovery time is infinitesimal with respect to the duration of half cycles of the ac wave. Therefore, the effect is negligible. The effect is not negligible at higher frequencies where reverse recovery time is an appreciable fraction of the half cycle of the ac wave.

The relatively high reverse-recovery time is the reason for the distorted rectification wave at 25 kHz. It is obvious that rectification becomes more imperfect as frequency is increased. This imperfect rectification not only increases power dissipation within the diode itself, but imposes extra burdens on other solid-state devices. Capacitors are forced to carry greater ripple currents and harmonic energy contributes both to temperature rise and EMI. Loads intended to operate from smooth dc now are subjected to appreciable ac components. Power-supply regulation, operating efficiency, and other performance parameters can be seriously degraded by this rather subtle behavior of inappropriately selected diodes. Mysterious penetration of the SOA boundaries of switching transistors often results from the reverse-recovery inadequacy of the associated free-wheeling diode.

Although rectifying diodes have been singled out, all other diodes in high-frequency power circuits are candidates for trouble if their reverse-recovery charac-

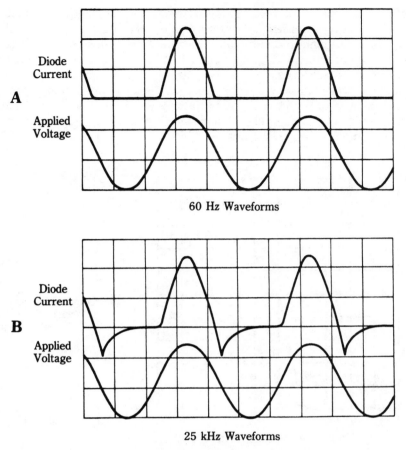

60 Hz Waveforms

25 kHz Waveforms

1-27 Effect of frequency on rectification by a PN junction diode.

teristics are too slow for the frequency being used. This includes free-wheeling diodes, circuit-isolating diodes, clamp diodes, charging diodes, commutation diodes, protection diodes, etc.

The remedy for such potential poor performance is to consult manufacturer's specifications with respect to the recommendation of diodes at the frequency of interest. For PN junction diodes, this task becomes increasingly difficult as frequency increases. After 50 kHz or so, the trade offs pertaining to other diode characteristics, such as forward voltage drop, might no longer be acceptable.

Fortunately, the Schottky diode, unlike the PN junction diode has no minority charge storage associated with its operation. Here, *reverse recovery* can be said to occur in zero time. As would be expected, this preserves near-ideal rectification well into the several-hundred kHz region. Practical limitation of high-frequency use is eventually approached because of internal capacitance, but not because of lack of intrinsic switching speed. The Schottky diode is a low-voltage rectifier; if the impressed voltage is too high, appreciable reverse leakage current sets in.

This, of course impairs the rectification behavior and efficiency in a manner not unlike that resulting from reverse-recovery time in PN junction diodes. Although Schottky diode ratings are continually improved, care must be exercised when using these devices to handle greater than 40 V or so in the reverse direction. A nice feature of the Schottky diode is its relatively low forward voltage drop. (See Fig. 1-28.) This vastly improves rectification efficiency in 5-volt power supplies. It also allows the use of full-wave rectifier circuits with lower loss than would occur from the use of PN junction diodes. The dashed curve of Fig. 1-28 depicts the penalty paid in some fast-recovery types.

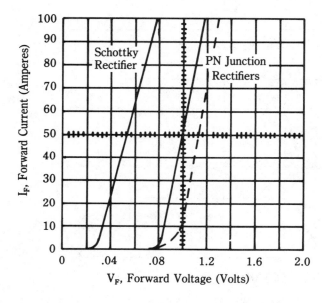

1-28 Comparison between typical 100 A Schottky and PN junction diodes.

The bottom line is to use Schottky diodes where voltage, temperature, and cost considerations allow in high-frequency power circuits. Otherwise, obtain appropriate fast-recovery PN junction diodes. Sometimes, too abrupt a reverse-recovery characteristic provokes ringing in circuit and wiring inductances and in excessive RFI. In such instances, make a compromise in the selection of recovery characteristics.

The thermal circuit

One of the basic requirements of designing power-control systems is to operate the power-handling devices at safe temperature levels. This calls for adequate heat removal, which can be achieved via simple calculations involving the relevant thermal parameters. These are power, P thermal resistance, Θ, and temperature, T. Fortunately, you can set up Ohm's law relationships between pairs of these parameters. This can be readily grasped by referring to the thermal circuit of Fig. 1-29.

$$T_J > T_C > T_S > T_A$$

Heat Input Is Represented by Power P

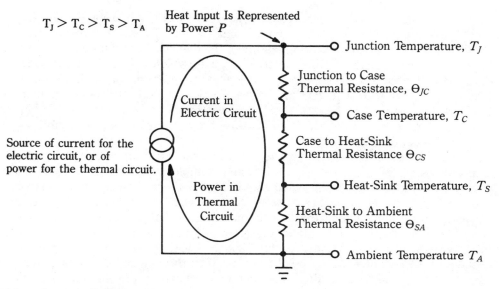

Junction Temperature, T_J

Junction to Case Thermal Resistance, Θ_{JC}

Current in Electric Circuit

Case Temperature, T_C

Source of current for the electric circuit, or of power for the thermal circuit.

Case to Heat-Sink Thermal Resistance Θ_{CS}

Power in Thermal Circuit

Heat-Sink Temperature, T_S

Heat-Sink to Ambient Thermal Resistance Θ_{SA}

Ambient Temperature T_A

Note: Symbols $R_{\Theta JC}$, $R_{\Theta CS}$, etc., are also used.

1-29 The thermal circuit—a contrived concept simulating Ohm's law relationships.

The basic relationships are:

$T = P \times \Theta$	(Temperature equals power times thermal resistance)	(1-1)
$P = T/\Theta$	(Power equals temperature divided by thermal resistance)	(1-2)
$\Theta = T/P$	(Thermal resistance equals temperature divided by power)	(1-3)

Note that these are simple algebraic relationships. So far it is easy enough to see that the thermal circuit is treated somewhat as an analogy of an electric circuit. Also, the parameters are not hard to obtain or measure. It happens that thermal power, P, can be measured electrically from the product of voltage and current; that is, electrical power in watts is taken also to represent thermal power. In many practical thermal problems, P also can be conveniently determined from equation (1-2) above. This is because Θ is often readily available from spec sheets, and temperature T can be determined with a thermometer or be postulated. (It can also be obtained from emitter-base voltage measurements.)

So far, so good. However, pay heed to unique aspects of the thermal circuit. First, the temperatures that might be determined from thermometer instruments, unlike voltage readings in the electric circuit, are not referenced to zero. Ground in the thermal circuit is ambient temperature, generally taken as 25°C. This must be kept in mind, for it is the temperature difference across any portion of the thermal circuit that must be used in the above equations for T.

Second, from what has been so far stated, the unit of thermal power is the watt. The unit of thermal voltage—that is, temperature—is the degree Celsius.

And the unit of thermal resistance, Θ, is degrees Celsius per watt, $C°/W$. Note that no direct use is made of a thermal current.

You now arrive at an interesting feature of the thermal circuit; it appears not to be a true analogy of the simple electric circuit to which it is likened. Consider equation (1-2) above. It is easy enough to liken temperature to voltage in the sense that increases in both give rise to increased power. In the electrical situation, power is proportional to E^2, but thermal power is only proportional to T. This seems to suggest that voltage and temperature are not truly analogous quantities. Where, indeed, lies the fallacy?

Actually, further attempts to make the electrical and thermal circuits appear analogous only lead to further frustration. For example, power takes the place of a parameter analogous to current when comparing the thermal circuit to the electrical circuit. You must conclude that the so-called thermal circuit has been contrived to yield relationships among the thermal parameters that can be calculated in a similar manner as electrical parameters in an electrical circuit. It turns out that this contrivance is exceedingly useful in a practical sense. Do not get upset because of the lack of a one-to-one analogy between thermal and electrical parameters. The practicality of the thermal-circuit concept can best be demonstrated by an example:

> Assume a power transistor is used as a series-pass element in a regulated power supply. Suppose that maximum power dissipation takes place when this transistor passes 3 A (amperes) and develops an accompanying voltage drop of 4 V. The resultant 12 W of electrical-power dissipation also represents 12 W of thermal-power dissipation. A destructive rise in junction temperature will occur unless a heat-removal technique is used. Heat removal is needed because the specifications for the transistor indicate a junction temperature exceeding the maximum allowable level of 200°C if the transistor were to be operated in still air with 12 W of dissipation and no heat sink. In order to select an appropriate heat sink, make use of the thermal-circuit equations and procede as follows:
>
> 1. Θ_{JC} and T_J are obtained from the manufacturer's specifications of the power transistor. A typical power-transistor spec sheet is shown in Table 1-4. Although this spec sheet does not pertain to the example at hand, it does demonstrate the data obtainable in such manufacturers' literature. Although a T_J of 200°C is often cited, a street-wise safety factor is used, so 175°C is assumed in the example. At this time, record also the value of Θ_{JC}, which will be assumed to be 6°C/W.
> 2. Next use the thermal Ohm's law relationship:

$$\Theta_{JA} = \frac{T_J - T_A}{P}$$
$$= \frac{175 - 25}{12}$$
$$= 12.5°C/W$$

**Table 1-4. Typical spec sheet depicting electrical and thermal data
on power transistors.**

Rating (Maximum)*	Symbol	2N6544	2N6545	Unit
Collector-Emitter Voltage	$V_{CEO(sus)}$	300	400	Vdc
Collector-Emitter Voltage	$V_{CEX(sus)}$	350	450	Vdc
Collector-Emitter Voltage	V_{CEV}	650	850	Vdc
Emitter Base Voltage	V_{EB}	9.0		Vdc
Collector Current—Continuous	I_C	8.0		Adc
—Peak (1)	I_{CM}	16		
Base Current—Continuous	I_B	8.0		Adc
—Peak (1)	I_{BM}	16		
Emitter Current—Continuous	I_E	16		Adc
—Peak (1)	I_{EM}	32		
Total Power Dissipation @ T_C = 25°C	P_D	125		Watts
@ T_C = 100°C		71.5		
Derate above 25°C		0.714		W/°C
Operating and Storage Junction Temperature Range	T_J, T_{stg}	−65 to +200		°C

Thermal Characteristic	Symbol	Max	Unit
Thermal Resistance, Junction to Case	$R_{\theta JC}$	1.4	°C/W
Maximum lead Temperature for Soldering Purposes: 1/8″ from Case for 5 Seconds	T_L	275	°C

*Indicates EDEC Registered Data
(1)Pulse Test: Pulse Width = 5 ms, Duty Cycle ≤ 10%

Note: The transistors depicted in this table are quite different from the transistor postulated in the example. The table does demonstrate the electrical and thermal data available from manufacturer's spec sheets.

Motorola Semiconductor Products, Inc.

You now have the total thermal resistance of the thermal circuit.

3. Note what is thus far accomplished. You have determined the overall resistance, Θ_{JA}, of the resistive network of Fig. 1-29, and the resistance, Θ_{JC}, of one of the three resistive elements. If you can determine just one more of the two remaining resistances, the remaining one can be obtained by simple arithmetic from the fact that the total resistance of a series circuit of resistances is the sum of the individual resistances. It turns out that Θ_{CS} can be conveniently found in the technical literature. An excerpt from a table supplying this information is shown in Table 1-5. Assuming the power transistor has a TO-66 package with a 2-mil mica insulating spacer, determine Θ_{CS} to be 2.3°C/W.

4. From:

$$\Theta_{SA} = \Theta_{JA} - \Theta_{JC} - \Theta_{CS}$$

You have:

$$\Theta_{SA} = 12.5 - 6.0 - 2.3 = 4.2°C/W$$

Table 1-5. Case-to-sink thermal resistance for different packaging and mounting methods.

		θCS		
		Metal-to-metal	**Using an insulator**	
		With heat-sink	**With heat-sink**	
Case	**Dry**	**compound**	**compound**	**Type**
TO-3			0.4°C/W	3 mil mica
TO-3	0.2°C/W	0.1°C/W	0.35°C/W	Anodized Aluminum
TO-66	1.5°C/W	0.5°C/W	2.3°C/W	2 mil mica

Motorola Semiconductor Products, Inc.

5. Finally, the selection of an appropriate heat sink again turns out to be a look-up procedure. Table 1-6 is typical catalog data depicting heat sinks suitable for various Θ_{SA} ranges. A Thermalloy 6141 would meet the heat-removal requirements of the power transistor operating under the stipulated conditions. Alternatively, there are other Thermalloy models, as well as those offered by Wakefield, IERC, and Staver. All of these are suggested for the 3.0 to 5.0°C/W thermal-resistance range.

The above example assumes natural convection; that is, no blowers or fans. When forced-air convection is used, it is equivalent to reducing Θ_{SA}. Appreciable improvement in heat removal is often attained, but calculation is generally not easy. You must dig up graphical and empirical data as guidance. A particular difficulty is encountered in assessing the nature of the moving air in the region near the

Table 1-6. Suitable heat sinks for various θ_{SA} ranges.

θSA*(°C/W)	TO-3 & TO-66 Manufacturer/Series or Part Number
0.3–1.0	Thermalloy—6441, 6443, 6450, 6470, 6560, 6590, 6660, 6690
1.0–3.0	Wakefield—641
	Thermalloy—6123, 6135, 6169, 6306, 6401, 6403, 6421, 6423, 6427, 6442, 6463, 6500
3.0–5.0	Wakefield—621, 623
	Thermalloy—6606, 6129, 6141, 6303
	IERC—HP
	Staver—V3-3-2
5.0–7.0	Wakefield—672
	Thermalloy—6002, 6003, 6004, 6005, 6052, 6053, 6054, 6176, 6301
	IERC—LB
	Staver—V3-5-2
7.0–10.0	Wakefield—672
	Thermalloy—6001, 6016, 6051, 6105, 6601
	IERC—LA, uP
	Staver—V1-3, V1-5, V3-3, V3-5, V3-7
10.0–25.0	Thermalloy—6013, 6014, 6015, 6103, 6104, 6105, 6117

Motorola Semiconductor Products, Inc.

cooling fins of the heat sink. Not only is the air velocity an important factor, but you should know whether the air motion is laminar or turbulent. Experience and experimentation certainly enter the picture here.

Conduction and convection are not the only mechanisms of heat removal. Radiation can be advantageously used because heat energy propagates into space in proportion to the fourth power of absolute temperature. To exploit this phenomenon, it is necessary to be aware of the nature of various surfaces. A commonly encountered fallacy is that mirror-like surfaces are best for enhancing cooling via radiation.

The logic advocated in support of this fallacy often cites the household electric heaters utilizing highly polished parabolic reflectors to direct the heat from the heating element into the room. This, however, is not the same thermal situation existing in heat sinks where the hot heat sink itself is the radiating element. Indeed, if the heat sink body and fins had mirrorlike surfaces, radiation efficiency would be negligible.

This is readily discernible from inspection of surface emissivities listed in Table 1-7. Note, for example the great difference in emissivity between polished and oxidized copper, or similarly, between polished and anodized aluminum. Physics textbooks reveal that surfaces that are good radiation absorbers are also good radiators. In general, rough, dark surfaces that are not like mirrors are the efficient radiation absorbers—and radiators.

**Table 1-7. The radiation efficiency
of various surfaces.**

Surface	Emissivity, E
Polished aluminum	0.05
Polished copper	0.07
Rolled sheet steel	0.66
Oxidized copper	0.70
Black anodized aluminum	0.7 –0.9
Black air drying enamel	0.85–0.91
Dark varnish	0.89–0.93
Black oil paint	0.92–0.96

Note: An emissivity of 1.00 corresponds to 100 percent radiation efficiency.

National Semiconductor Corp.

A common controversy involves the choice between aluminum or copper as heat sink material. If the comparison is made on the basis of density or weight, aluminum turns out to be the superior heat conductor. If, however, the comparison is made on the basis of cross sectional area, copper excels. Both metals are easy to fabricate in various shapes. Aluminum tends to be the popular choice of heat sink manufacturers, but it is well to keep in mind that for heat sinks of the same physical size and shape, copper will tend to outperform its aluminum counterpart. However, much also depends upon the overall nature of the thermal environment, such as the part played by radiation and convection. Other things being equal, aluminum tends to be the material of choice in terms of cost and availability.

The ceramic material, beryllium oxide, or beryllia, is often used to conduct heat from the semiconductor to its case because this excellent heat conductor is also an electrical insulator. This unique property also makes it useful as washers or spacers between power devices and heat sinks.

WARNING

Never abrade, cut, or drill this material—it is exceedingly hazardous to health in its powdered or pulverized form. A very tiny amount inhaled or ingested can be very toxic. It must not be disposed of by incineration or allowed to fall into the hands of unaware persons. But, you need have no fear of handling the substance in its solid form.

2
Solid-state amplifiers

POWER AMPLIFIERS CONFIGURED AROUND DISCRETE DEVICES AND FAIRLY
ordinary silicon output transistors have yielded surprisingly good performance;
yet, there has been room for improvements not readily achievable with the traditional devices. Consider, for example, the technique of negative feedback, one of
the important circuitry weapons for extending frequency response and ironing out
the effects of nonlinearities. The mere accomplishment of reasonably flat frequency response, say from 20 Hz to 20 kHz, never told the whole story with regard
to high-fidelity reproduction.

Missing from the response specifications was what was happening beyond 20
kHz. At first glance, this seems inconsequential because most people's hearing
tends to go into high attenuation after 17 kHz or so. However, when an audio
amplifier exhibits nonlinearity or positive feedback at higher frequencies, the
quick transient response of the amplifier is degraded; in addition, beat frequencies
and intermodulation distortion are sure to creep into the audio band itself. As a
rule of thumb, it would be wise to have audio amplifiers be flat out to about 200
kHz. But there is more to this requirement than meets the eye. Stereo equipment
manufacturers have long assumed that the shortcomings of the nonfeedback
amplifier could be overcome with sufficient feedback. The premise has been that if
a little feedback is good, more must naturally be better.

However, engineers in more exacting fields such as telephony or instrumentation have always been aware that it is best to start originally from the best nonfeedback amplifier achievable before piling on the negative feedback. Don't be too
severe with the stereo manufacturers; they really have done what was possible at a
reasonable cost. The power transistors that have been used in stereos have been
far from ideal devices for linearity and frequency response. Then, there has been
crossover distortion from the class AB push-pull output stages. As if this wasn't

enough, further departure from linearity came from the commonly used quasi complementary-symmetry circuits.

These and other shortcomings are overcome by using some of the newer power devices. And almost everything that has been said with regard to stereo amplifiers also applies to servo amplifiers, especially where high accuracy and low overshoot are performance goals. Oscillation, hunting, and other mechanical aberrations plague servo systems with too much patchwork negative feedback. Here again, some of the modern power devices tend to produce better results than many of the traditional brute-force power transistors.

Simple intercoms with audio power ICs

Power op amps, especially those designed for use in audio applications, are a natural for simple but effective intercommunications systems. Such intercoms are essential in industrial and commercial environments, but also find considerable residential use. Further simplifying practical implementation is the fact that the common dynamic speaker also functions exceedingly well as a microphone. The easiest intercom system to build, and often to install, is the master-slave format. In this scheme, the operator at the master station actuates a mode switch to determine when to talk or listen. The person at the slave, or remote station then talks when given the opportunity to do so, that is, when the mode switch at the master station is placed in its listen position.

Figure 2-1 shows two intercom systems utilizing audio ICs. All circuitry, including the master speaker/microphone, is located at the master station; the slave speaker/microphone is located at the slave or remote station and is not encumbered with any additional circuitry. In Fig. 2-1A, the LM388 IC can deliver in excess of 1.5 W to an 8 ohm speaker. A 10 to 12 V power supply is suitable. Class AB operation is used by the output stage of this IC so that the quiescent current is quite low, about 18 mA.

The 5 W intercom shown in Fig. 2-1B is actually more typical of such single IC circuits. It is similar to that of Fig. 2-1A, but a step-up transformer is inserted in the input circuit when either station uses the speaker as a microphone in the talk mode. This transformer is needed because of the lower gain of the more powerful ICs. A 5 or 6 V filament transformer will probably serve this purpose if operation is limited to voice transmission. The LM384 IC will operate well from a nominal 25 V dc supply. Approximately four or five square inches of copper foil on a PC board will function as a satisfactory heat sink.

Both of the intercoms just discussed will work with 4-, 8-, or 16-ohm speakers. More available power but lower power gain will be forthcoming from the lower-impedance speakers. However, higher-impedance speakers serve as more sensitive microphones. Speakers with 8 ohm impedance should prove to be a good compromise. If 4 ohm speakers are used with the LM384, a clip-on heat sink might be necessary. A relatively large-diameter speaker best serves the acoustical situation likely to prevail at the remote station; at the master station, the operator is close for both talking and listening, so a small speaker should prove suitable.

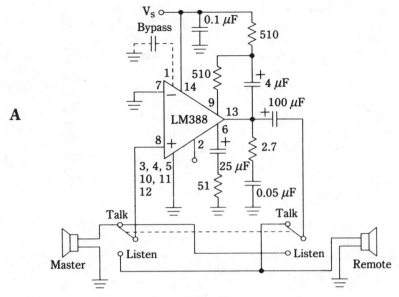

A 1.5 W master-slave intercom.

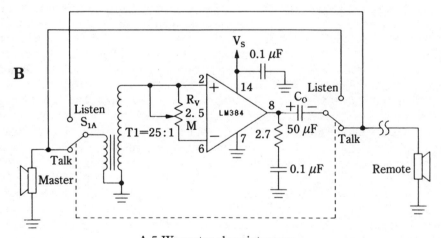

A 5 W master-slave intercom.

2-1 Simple intercom systems using power ICs. National Semiconductor Corp.

The optoisolator as an
interstage coupling element

Optoisolators are very useful devices for many circuit applications. They are often found in feedback loops of switching power supplies and have found extensive application in the transmission and processing of digital data. They also serve well where it is desirable to isolate control-panel switches and control knobs from the lethal high-voltage systems being controlled. In these applications, they are often found to be less costly and less demanding of space than are transformers. Optoisolators commonly employ an LED (light-emitting diode) infrared emitter and a variety of photosensitive detector elements, including photodiodes, phototransistors, photo Darlingtons, photo-FETs, and photothyristors. High-voltage isolation capability stems from the physical separation between emitter and detector. Notwithstanding their widespread uses, relatively little has been done with these devices for application as a linear transducer or coupling agent. At first thought it would appear that the optocoupler would make an ideal interstage coupler in audio amplifier systems. This would resolve conflicts in bias and dc voltage levels and would be expected to exhibit flatter frequency response than either transformers or capacitors.

Qualitatively, this line of reasoning points in the right direction. However, practical implementation has been held in check by problems in linearity, temperature dependence, and a general transfer function that has not impressed designers as being sufficiently predictable. This unpredictability has been somewhat frustrating, inasmuch as the optoisolator ability to interrupt ground loops can sometimes be very useful. For example, in intercom systems it sometimes happens that considerable ac hum is picked up because of the way industrial and commercial utility installations distribute power within a building. It fortunately turns out that if passage of dc levels is not required, the optoisolator can be made to provide sufficient linearity for some audio-coupling applications, such as intercoms and voice-processing sections of amateur-radio transmitters. The secret of such implementations is to operate within a relatively linear region of the optoisolator transfer characteristics, and to stabilize the unmodulated LED current.

The circuit shown in Fig. 2-2 provides good linearity and a reasonable current-transfer ratio. The 10 mA constant-current source need not necessarily be a two-terminal device. For example, the single-transistor current regulator depicted in Fig. 1-3C is suitable. Many three-terminal voltage-regulator ICs can be used as constant-current sources. Also, it might not be necessary in all applications to use separate 5 V dc supplies and separate ground systems.

An alternative optoisolator scheme for the interstage coupling of audio signals is shown in Fig. 2-3. Here, current stabilization in the LED circuit is achieved via a differential amplifier. Although no dc blocking capacitors are shown, their use will be called for in many applications. Otherwise, this circuit operates in much the same manner as the circuit in Fig. 2-2. The unmodulated LED current is governed by the resistor labeled bias. Empirical determination of this resistor is necessary for the attainment of optimum linearity.

When either of these circuits is used without the dc blocking capacitors, the low frequency response extends all the way down to zero (dc). However, it is much more difficult to obtain proportional transfer of the dc component than the ac portion of the signal, and temperature effects then tend to make operation somewhat erratic.

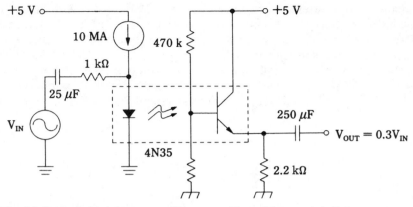

2-2   Method of obtaining a near-linear coupling with an optoisolator. General Electric Co.

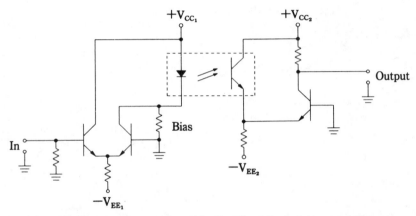

2-3   Alternative scheme for coupling audio signal with an optoisolator.

Getting the best from MOSFET power

The high-impedance input characteristic of the power MOSFET has excited a great many designers, because this device also possesses high frequency capability. At dc and audio frequencies, easy drive and flat response are, indeed, readily realized. Realize that at higher frequencies this device has a high input capacitance, and the effect of this can be made worse by the Miller effect. Thus, at higher frequencies, the power MOSFET no longer appears to be a near-infinite input impedance device. Its frequency response will roll off long before the actual amplifying capability of the device has suffered. The early frequency roll-off actually takes place in the driver output circuit.

What is needed is a very low-impedance driver. With such a driver, the input capacitance of the MOSFET will have minimal negative effect, and wideband performance can be attained. If you can devise a low-impedance drive circuit and still retain the high impedance input characteristic of MOSFET devices, you will have a very useful amplifying system. This concept is particularly true if such performance can be coupled with the high output capability of the power MOSFET. These statements might initially appear to involve contradictions, but the circuit shown in Fig. 2-4 readily provides all of the desirable operating parameters: high-impedance input, wideband frequency response, and high power output.

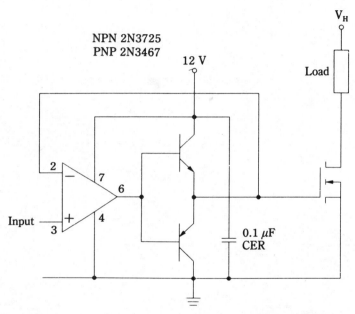

2-4   Driver circuit allows high-frequency response and high input impedance from MOSFET amplifier. International Rectifier Corp.

In Fig. 2-4, the input device is an operational amplifier with a MOSFET input stage. The input capacitance is approximately 4 pF (picofarads) compared to 700 pF or so for a power MOSFET. The intermediate stage is a complementary-symmetry, emitter-follower circuit for further reducing the driving impedance seen by the power MOSFET stage. Because of the feedback connection, the first two stages perform as a voltage follower. This feedback technique makes the output impedance of the complementary-symmetry stage lower than it would otherwise be. So in effect, the power MOSFET output stage is driven by an exceedingly low-impedance source, and can thereby deliver wideband frequency response. Depending somewhat on the selection of the bipolar driving transistors and the power MOSFET, an 8 to 12 MHz passband can be realized with this scheme. Of course, the load must have suitable wideband response in order to achieve an overall flat response from the amplifying system.

A TV sound-channel IC

Workers in the power-control field have become accustomed to the trend to accommodate both high- and low-power devices on a single chip. The Darlington power amplifier was the first example of monolithic integration of driver and output transistor in an IC module. Transistor arrays and more complex integrations followed, often involving several stages, together with auxiliary stages, such as voltage regulators, protective circuits, etc. A classic example was the three-terminal voltage regulator. As sophisticated as some of these ICs were, they generally did not process widely different kinds of signals, such as those encountered in radio and TV circuits. It was long felt that a diverse mix of functions on a single chip would lead to compromised performance, and a tendency towards various types of instabilities from inadvertent cross talk and unintended feedback paths.

This is no longer the case, as can be witnessed by inspection of late-model TV sets. In place of the numerous discrete stages and high parts count, you now see a relatively small number of multiple-function ICs and a dramatic reduction in associated parts. An example of such an IC is the Sprague ULN-2290B TV sound channel. In addition to an audio-output stage with 4 W capability, this IC contains the audio preamplifier, a regulated power supply, FM detector, and a six-stage limiting IF (intermediate frequency) amplifier. It also contains or has pinout provisions for important circuit functions, such as low-pass filtering following IF amplification, deemphasis, IF amplifier bias, audio feedback, and volume control. The input impedance of the IF amplifier is sufficiently high to enable optimum use of either a ceramic filter or a tuned circuit. The functional block diagram of this IC is shown in Fig. 2-5.

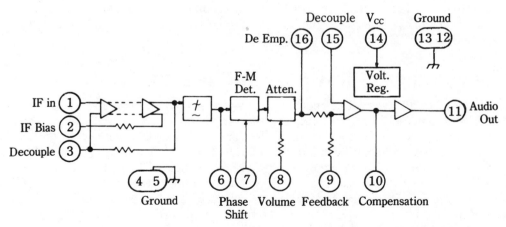

2-5   Block diagram of the ULN-2290B TV sound channel IC. Sprague Electric Co.

The way this IC is used in a TV receiver is shown in Fig. 2-6. Note that this device, in association with a relatively small number of external components, dramatically simplifies the circuit layout of a TV set. When used in conjunction with

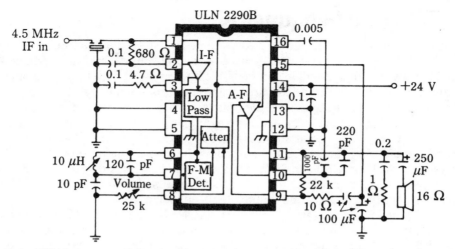

2-6   TV implementation of Sprague ULN-2290B TV sound channel IC. Sprague Electric Co.

other dedicated ICs, the overall savings in assembly labor provides the consumer with much more performance per unit cost than was hitherto attainable. Note that connection terminals 4 and 5, as well as 12 and 13, are at ground potential and are in the form of tabs. These tabs can be conveniently provided with simple heat sinks.

ICs of this type have already given a good account of themselves with regard to performance and reliability. However, considerable trouble has been experienced with the sockets. It behooves the designer to carefully consider the mechanical integrity of sockets, and to focus attention on contact materials with good resistance to corrosion; otherwise, the multifunctional IC is not given a fair chance to perform within its capabilities.

Dedicated IC low-cost audio systems

Millions of small AM radios, phonographs, intercoms, and other audio equipment have been manufactured with output power levels in the several-watt range. For the large but nondemanding sector of the consumer market, such output power has proven to be adequate. Among the multitudes of satisfied consumers are TV viewers. Often an otherwise elaborate and sophisticated TV chassis will be equipped with a tiny speaker driven by an audio output stage with a couple of watts capability. Of course, the prime explanation of this hinges on cost and competition. But minimal power audio can be satisfactory, particularly if the quality of sound reproduction is good.

The Sprague ULN-2280B is a monolithically integrated 2.5 W audio system requiring a minimal number of external passive components. This IC contains 23 transistors. The savings in expense, construction effort, and debugging time are compelling reasons for choosing such an IC in place of the extensive parts count of discrete devices and supporting circuitry.

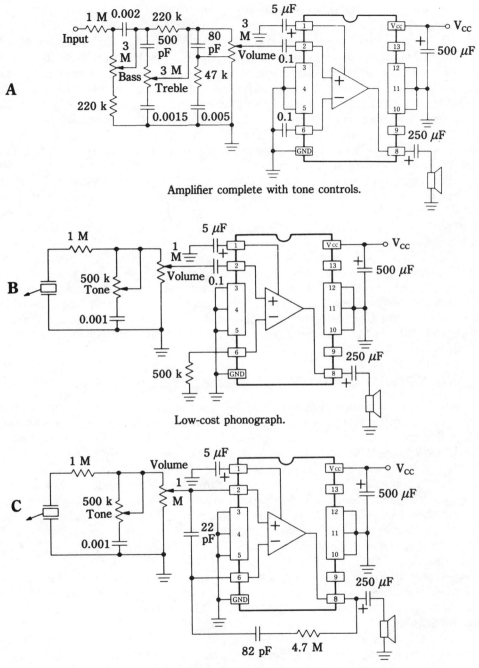

Amplifier complete with tone controls.

Low-cost phonograph.

Amplifier with base-boost network.

2-7 Three easily implemented audio applications of the ULN-228 IC. Sprague Electric Co.

Figure 2-7 depicts three applications of this audio IC. To assure satisfaction with their products, manufacturers have exerted efforts beyond the mere placing of such sophisticated devices on the market. Additionally, they have developed peripheral circuitry to bring out the best possible results from their ICs. This relieves the user of unnecessary design drudgery, as well as experimentation that could result in accidental damage. The three audio applications shown all use input tone controls. Base boost is a useful expedient for audio systems with small speakers, because partial compensation for the early low-frequency roll-off of the small speaker is thereby attained. (The hi-fi enthusiast will surely object to such an approach, but such elite hobbyists would hardly be expected to be interested in a 2.5 W audio amplifier in the first place.)

For 8 ohm speaker operation, the prescribed power supply voltage is about 17 V. For 16 ohm speakers, the power supply should develop 22 V. In both instances, total harmonic distortion will be close to 3 percent. Somewhat more power can be extracted if 10 percent of THD (total harmonic distortion) is permissible (as it might be for voice only).

The ULN-2280B has a high-impedance input and a fixed voltage gain of 34 dB. An inexpensive heat sink suffices, the recommended one being a Staver V-8, which clips neatly onto the IC package. Thermal overload protection is incorporated in the internal circuit so that output current is temperature limited. The IC is also self-protected against short circuiting of the ac output. All things considered, dedicated ICs of this type are excellent devices for the economy-minded hobbyist bent on achieving rapid and predictable results.

Two stereo channels on a chip

Another low-power audio amplifier is of interest because it closely approaches the ideal system concept—the system on a chip alluded to in chapter 1. The stereo-amplifier circuit shown in Fig. 2-8 is actually a single IC, the National Semiconductor LM379. This IC is in essence a dual-power op amp. It is representative of the modern trend to combine signal processing and high-power capability in a single module. Whereas 6 W per channel wouldn't impress certain audio enthusiasts as high power, those acquainted with several hundred milliwatt monolithic ICs will accord the LM379 IC a healthy respect. This power level is very satisfactory for stereo phonographs, AM-FM radio receivers, tape recorders and players, and intercoms. It is also quite possible to upgrade the sound system in many TV sets with this simple audio amplifier.

As seen in Fig. 2-8, it is only necessary to associate about eighteen passive components, two speakers, and a single power supply with this IC. Also, it should have adequate heat sinking. Heat sinking might be provided by the chassis mounting provision.

This amplifier is given a bass-boost response by the 0.02 μF capacitors in the feedback loops. As frequency decreases, the reactances of these capacitors increase; this in turn develops less negative feedback which results in greater gain at the lower frequencies. The intent is to compensate for the poor low-frequency response of small speakers.

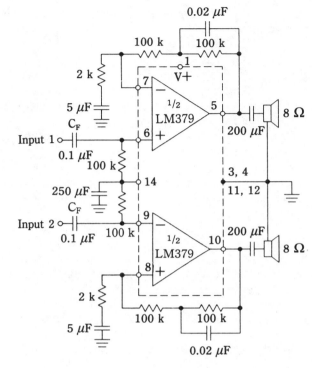

2-8 A 6 W per channel stereo amplifier using a single dedicated IC.
National Semiconductor Corp.

A simple 24 V unregulated power supply will suffice. Somewhat improved performance will result from the use of a regulated dc supply. This IC has internal current limiting and thermal protection. A nominal 70 dB of channel separation exists despite the physical proximity of the circuits on the chip. Internal compensation and about 90 dB of open-loop gain enables stable operation with the negative-feedback paths in place. It is obvious that LM379 is a dedicated IC intended for hi-fi audio applications. This will likely lead to better results than the mere adaptation of a general-purpose power IC.

The experimentally inclined might perceive this setup as a convenient means of driving higher-power output stages. In such an arrangement, a negative-feedback loop should be connected around each channel. If this is done while recording distortion, a little cut and try will lead to optimum overall performance. The easiest way to drive a more powerful output stage is via ac, that is, capacitive coupling.

Bridge amplifiers

Most audio amplifiers have either single-ended or push-pull output stages. Bridge-connected output stages comprising four active devices are known to have many desirable features, but the extra complication involved is generally not attractive. In other power-control applications, such as motor control and switching power supplies, the bridge format is often used, particularly where high power levels must be efficiently processed. There is, however, a low to moderate power application in audio amplifiers where the bridge configuration can be advantageously used in a simple manner.

Most power ICs have push-pull output stages. If two such ICs are used so that the single load is connected between their outputs, the load will see a bridge amplifier, providing the amplifier inputs are appropriately driven. Thus, the system will require only two rather than four, power devices from the hardware standpoint; even better, in some cases only a single IC will be needed because some power ICs are duals. If the power ICs are dedicated audio circuits, there will be very little effort needed to quickly assemble a quality performance amplification system because the important engineering and design techniques are already incorporated within the ICs. Incidentally, the bridge format is a neat way to use direct coupling to the speaker without the need for a split power supply or a large electrolytic coupling capacitor.

Three bridge-amplifier circuits are shown in Fig. 2-9. The common denominator in these circuits is the connection of the load or speaker between the output terminals of the power ICs. Careful inspection will show that their input connections are not the same, however. This need not lead to confusion if the basic objective of operating the output stages is kept in mind. It is simply that from the load the outputs must appear to be series-aiding, that is, when one output is becoming increasingly positive, the other output should be, in mirror fashion, becoming increasingly negative. You might very well say that the outputs are in push-pull. The fact that each output stage is already a push-pull circuit makes the resultant circuit a bridge. The load experiences twice the voltage it would be impressed with from just one of these push-pull output stages. It is obvious that providing the dc operating voltage is doubled, the bridge arrangement doubles available output power.

In the circuits of Fig. 2-9, note that the voltage gains of the amplifiers are set by feedback networks at 100. It is also true that the output of the left-hand amplifier is reduced potentiometrically by a factor of 100 before application to the inverting input of the right-hand amplifier. Thus both amplifiers receive equal input voltages. Phasing is taken care of by the fact that the right-hand amplifier operates as an inverter. Actually, there is not perfect symmetry between the two amplifiers; the 100 kohms potentiometer is a balance adjustment—output current should be zeroed under quiescent conditions. The experimenter can add a volume control to the input circuit.

Audio power ICs for vehicular use

Power ICs for audio service in vehicles require special attention at the design and manufacturing stages. To begin with, the fact that such ICs must operate from nominal 12 V systems aggravates the already difficult task of handling a respectable power level in a monolithic module. For a given power rating, the 12 V IC must deal with higher currents than an IC specified for, say 20 or 30 V operation. Worst of all is the relatively hostile environment these ICs must work in. Tempera-

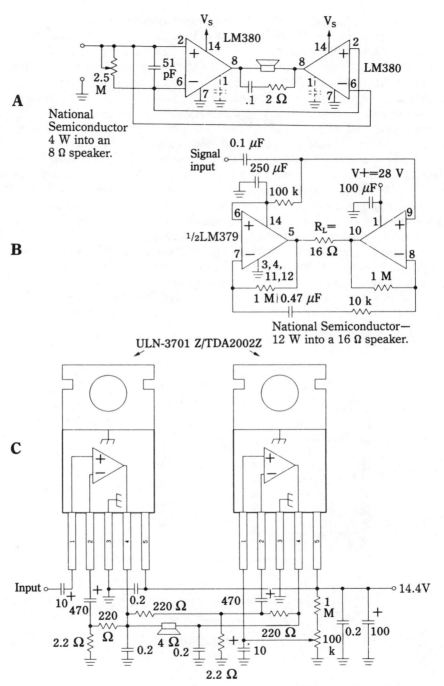

A

National
Semiconductor
4 W into an
8 Ω speaker.

B

C

ULN-3701 Z/TDA2002Z

Sprague Electric—115 W into a 4 Ω speaker.

2-9   Three circuits that present full-bridge output stages to the load.

ture swings are greater than that ordinarily experienced by semiconductor devices which perform in indoor equipment. Not only is the dc voltage from the battery alternator system quite variable, but it is burdened with transients that not only create interference, but are potentially destructive. Abusive conditions must be anticipated during equipment installation, battery replacement, and jump-starting. In order to merit selection over discrete-device audio circuits, the IC must provide output power in the 5 to 10 W range, must equal or outperform the distortion and frequency response readily attained with discrete designs, and must exhibit compelling advantages in simplicity and cost. Two such ICs are discussed.

The Sprague ULN-3703Z/TDA2003 IC power amplifier is shown in a typical circuit in Fig. 2-10. The speaker is a typical automotive type with a 3- to 4-ohm voice coil. This IC can also work into lower-impedance speaker loads. About 80 dB of open-loop gain is developed; in most applications, negative feedback results in a closed-loop gain of 40 dB. Suitable values of the feedback elements needed to bring this about are 39 ohms for R_{fb}, and 0.039 μF, for C_{fb}. The tab of the TO-220 plastic package is at ground potential—no insulation is required for mounting to the chassis. A nominal 6 or 7 W are readily obtained; with a 2-ohm load, the power output can be as high as 10 W. The output devices of this IC operate essentially in class B. Total harmonic distortion does not exceed 0.15 percent over most of the power and frequency range. The IC is internally protected against thermal over-

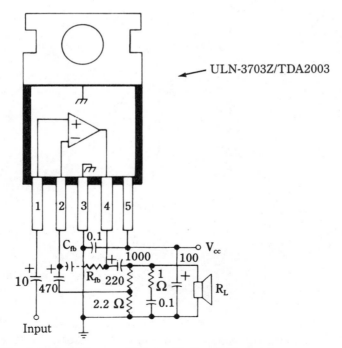

2-10 A dedicated power IC for vehicular audio service. Sprague Electric Co.

load, output overload, supply transients, ac/dc short circuits, polarity inversion, and open grounds.

Although the circuit depicts a single speaker, operating specifications are given more realistically for two speakers in parallel, this being the format most often encountered in modern vehicles. Thus the impedance load for two 4-ohm speakers would be 2 ohms, and for two 3.2-ohm speakers it would be 1.6 ohms. Before wiring speakers, be sure that they are phased so that the speaker cones move forward and backward in unison. This precaution is more relevant for stereo then binaural systems, but attention to many little details adds up to more pleasing audio reproduction.

For those interested in higher-power capability and stereo-type delivery, two-channel audio ICs are also available. An example is the Sprague ULX-377W, a dual-power amplifier of 10 W output per channel.

Another interesting audio power IC designed for the harsh vehicular environment is the RCA CA810Q/TBA810S. This device works optimally when driving a single 4-ohm speaker, under which condition it will typically deliver 6 W. It is particularly adapted to the needs of mobile communications and CB equipment. The output stage operates in class B as attested by a quiescent current drain of about 15 mA for the entire IC. Because of matched transistor characteristics and accurate biasing, crossover distortion is negligible. Total harmonic distortion is in the vicinity of 0.3 percent for output at the 3 W level. Open-loop gain is in the vicinity of 80 dB. Design flexibility is aided by the inordinately high input resistance of several megohms.

An audio amplifier using this IC is shown in Fig. 2-11. The values of C1 and R1 in the negative-feedback loop are selected to provide a closed-loop gain of 37 dB. Two unusual pinouts on this IC are pins 6 and 9. Pin 6 is a bootstrap provision for aiding the dynamic balance in the two output transistors. Pin 9, when con-

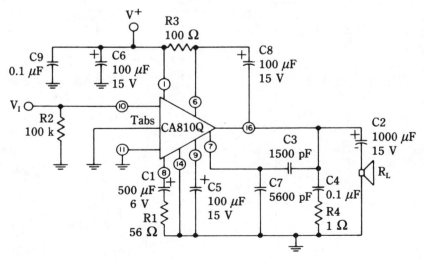

2-11 Another vehicular audio system using a dedicated-power IC. RCA

nected through a large capacitor to ground helps to immunize the amplifier against the effects of power supply variation. This reduces 120 Hz ripple from power-line rectifiers, and in automotive use, reduces the likelihood of alternator whine getting into the speaker.

To provide further insight into the nature of this power IC, the schematic diagram is shown in Fig. 2-12. It is obvious that a discrete-device replica of such a circuit would be expensive, and because of the direct coupling and the high overall gain, would be tricky to balance and stabilize; it also would be difficult to simulate the quick-acting thermal feedback which protects the output stage at precise overload limits. Figure 2-13 depicts the shutdown characteristics of this IC. The two curves reveal essentially similar information; one shows the protection in terms of output power, and the other curve shows the shutdown function in terms of current consumption. A nice feature of this protective technique is that normal operation automatically resumes when the abusive conditions relax sufficiently to allow the case temperature to return below 125 degrees Celsius. Note that between 125 and 150 degrees Celsius the IC remains operative, but at reduced current and power output capability. For the consumer market, such limiting action is more desirable than an abrupt turn-off action.

Heat removal is through two ground-potential tabs, which enable the PC foil to act as a heat sink. However, free air circulation is also required in most installa-

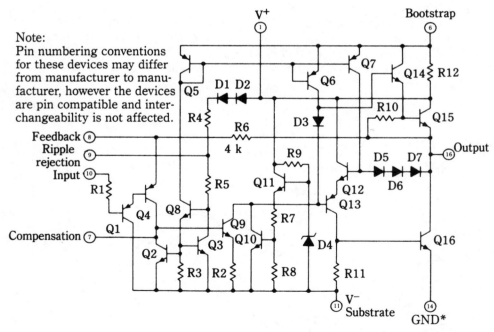

*Wing tabs are to be grounded.

2-12   Schematic diagram of RCA CA810Q/TBA810S audio-power IC. RCA

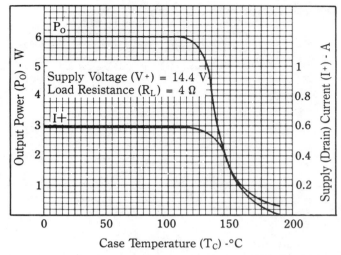

2-13 Thermal and overload shutdown characteristics of the CA810Q power IC. RCA

tions. Actually, two dual-inline 16-pin packages are available—one has straight tabs, coplanar with the package; the other has right-angle tabs for convenient PC board insertion.

Complementary-symmetry channel amplifier

A 1 to 3 W audio amplifier can provide satisfactory volume for listening purposes. Yet, as is well known, the tendency has been to use more powerful amplifiers, up to several hundred watts. The advantage of the amplifier with greater power capability is that it exhibits a greater dynamic range because it can handle loud passages without clipping the peak waveforms. (In practice, this might be partially offset in some cases because unless well designed, the powerful amplifier can suffer more from the effects of nonlinearity and crossover distortion at room volume.) Power is quite costly, and is generally attended by increased circuit complexity, diminished reliability, and inconveniences associated with heat removal, size, and weight. In practice, a reasonable compromise can often be realized at moderate power levels; 10 to 15 W might indeed sound as hi-fi as much higher power levels if other performance parameters are good.

The 12 W amplifier shown in Fig. 2-14 is relatively simple because it utilizes an IC input amplifier, needs no driver stages or elaborate protection circuitry, and makes use of a direct-coupled, complementary-symmetry output stage. Note that this amplifier is intended for use with an 8 ohm speaker. The output transistors are made specifically for service in complementary-symmetry amplifiers and are well matched in their important parameters. This yields better results than the practice of finding NPN and PNP power transistors capable of handling the requisite power

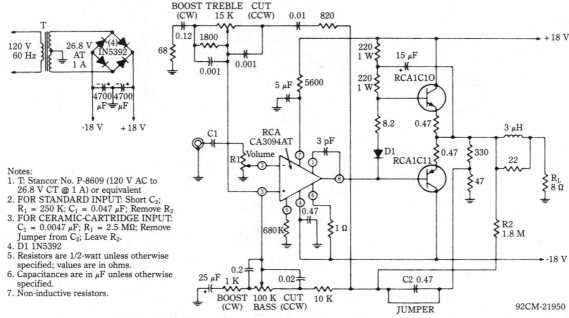

2-14 A 12 W complementary-symmetry audio amplifier. RCA

but often exhibiting appreciably different current gains, bias characteristics, and thermal parameters.

The integrated-circuit input amplifier facilitates implementation of treble and bass controls which are often lacking in much more powerful amplifiers. Also, special input circuit data is indicated for different input sources. In this regard, note that capacitor C1 is not merely an ac coupling capacitor, but is selected to provide appropriate frequency response. Capacitor C2 and resistor R2 are in the feedback path. They are also used to shape the frequency response to the characteristics of the source supplying the input signal. They are to be shorted or left out as indicated in the circuit notes.

The dual-polarity power supply has the topography of a bridge rectifier. However, because of the center tap on the power transformer, the supply actually operates as two separate full-wave sections, each section utilizing a pair of rectifying diodes. This has become common practice. It not only conveniently allows the dual-polarity format, but it provides full-wave rectification without the added diode voltage drop of a true bridge circuit.

A 20 W workhorse audio amplifier

The 20 W audio amplifier shown in Fig. 2-15 is tailor-made for the experimenter and hobbyist. It uses the minimal number of discrete transistors needed to provide acceptable performance. Because of the quasi complementary-symmetry output

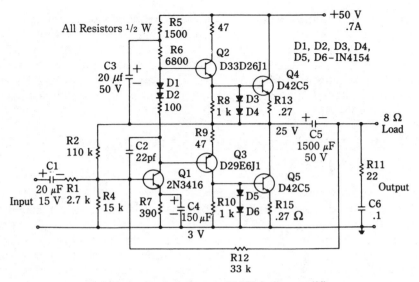

2-15 A simple low-cost 20 W audio amplifier.

stage, the two power transistors are identical NPN types featuring plastic encapsulation in the low-cost, tab-mount TO-220AB package. One single polarity 50 V power supply with 700 mA capability suffices to operate this amplifier. All of the resistors can be ¹/₂ W types, and all the diodes are identical types.

Capacitor C3 provides a small amount of positive feedback to equalize the positive and negative voltage swings at the output. This is a refinement to help make the output stage live up to its role as a symmetrical amplifier. Such localized positive feedback can be used to manipulate impedance levels and stage gains, but it does not confer any instability to the amplifier because overall feedback is always negative over the whole band of frequencies where gain exceeds unity. The main ac feedback path is through resistor R12. Overall feedback is prevented from being positive at higher-than-audio frequencies by the RC (resistive-capacitive) network, R11 and C6, connected across the output.

A common mistake made in the construction of quasi complementary-symmetry amplifiers is transposition of the driver transistors, one of which is an NPN type, the other a PNP type. In this circuit, Q3 is the PNP driver transistor. Because of the way it is used, power transistor Q5 acts as if it too was a PNP type.

Output transistors Q4 and Q5 are protected from overdrive by diodes D3 through D6. Because of the selected values of R13 and R15, these diodes conduct heavily and cause a drive limiting action very soon after the development of 20 W into an 8 ohm load.

Diodes D1 and D2 produce a voltage drop that biases the drivers; consequently the output transistors operate into the class AB region to limit crossover distortion to an acceptable level. In this amplifier, there is adequate temperature tracking between this bias source and the bias requirements of the output transistors so that idling current of the output stage remains reasonably constant over the ambient temperature range likely to be encountered in ordinary room operation.

Because of resistor and device tolerances, it might be desirable to empirically determine the exact value of R2 so that the power supply voltage splits equally between the two output transistors. Also, R1 should not be deleted because it causes a slight attenuation of the input signal; it is used to isolate the feedback circuitry from the signal source.

A 40 W complementary-symmetry stereo amplifier

The commonplace quasi complementary-symmetry audio amplifiers naturally evoke curiosity about the real thing. Part of the popularity of the quasi circuitry goes back to an earlier era of solid-state technology when the available power device was the germanium PNP transistor. However, this popularity was enhanced by silicon technology because for a long time there were only NPN power transistors, and the first silicon PNP types were inferior in ratings and reliability. That, of course, is no longer the case; indeed there are many NPN-PNP pairs made specifically for complementary-symmetry output stages. Even better, there are NPN-PNP Darlington transistors also made for this purpose. Because of their relatively high current gain, Darlingtons generally dispense with two driver stages in stereo amplifier formats. This not only simplifies the circuitry and reduces cost, but helps achieve better operation and greater overall reliability.

The RCA1B07 and the RCA1B08 are NPN and PNP Darlingtons, respectively. They are specifically specified as output devices in complementary-symmetry audio amplifiers. The push-pull circuitry using these devices attains a closer

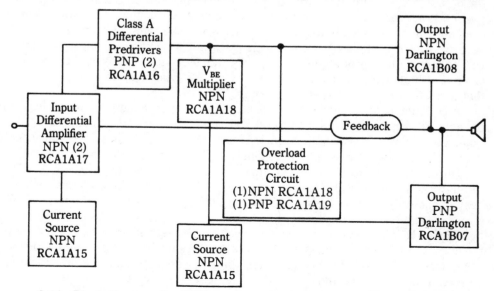

2-16   Block diagram of a 40 W complementary-symmetry audio amplifier. RCA

match between the half sections than can readily ensue from quasi-complementary circuits.

The block diagram of a 40 W audio amplifier for stereo systems is shown in Fig. 2-16. This block diagram relates the various transistors to their functions in the schematic diagram of Fig. 2-17. Note the several auxiliary functions not directly involved in amplification or in power boosting, per se. These are current sources, an overload-protection circuit, and a V_{BE} multiplier. The latter circuit provides operating bias for the output stages, enabling them to maintain their adjusted class AB mode over a wide temperature range. The overload-protection circuit senses output currents in the two halves of the output stage and limits any tendency towards dangerous increases in output by reducing the drives to the Darlingtons.

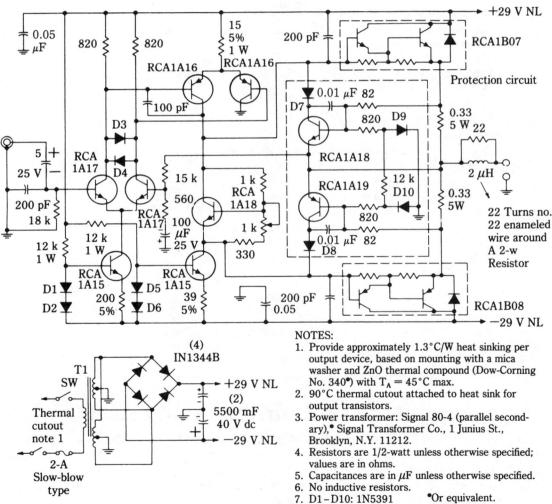

NOTES:
1. Provide approximately 1.3°C/W heat sinking per output device, based on mounting with a mica washer and ZnO thermal compound (Dow-Corning No. 340*) with $T_A = 45°C$ max.
2. 90°C thermal cutout attached to heat sink for output transistors.
3. Power transformer: Signal 80-4 (parallel secondary),* Signal Transformer Co., 1 Junius St., Brooklyn, N.Y. 11212.
4. Resistors are 1/2-watt unless otherwise specified; values are in ohms.
5. Capacitances are in μF unless otherwise specified.
6. No inductive resistors.
7. D1–D10: 1N5391   *Or equivalent.

2-17 Schematic diagram of a 40 W complementary-symmetry audio amplifier. RCA

The power supply does not operate as a bridge rectifier, in spite of its appearance. Rather, it comprises two full-wave, center-tapped rectifier circuits, one for each polarity. Note that the two identical windings of the power transformer are connected in parallel.

The performance data is in Table 2-1. Exceptionally wide frequency response is attained. Distortion is low because the phase and gain before feedback (that is, open-loop measurements) are quite good. It is seen that, although this amplifier is designed for 40 W output into a 4-ohm speaker, 30 W can be developed into an 8-ohm speaker.

Table 2-1. Performance data for a 40 W complementary-symmetry audio amplifier.

Measured at a line voltage of 120 V, $T_A = 25°C$, and a frequency of 1 kHz, unless otherwise specified.

Power:	
Rated power (4-Ω load, at rated distortion)	40 W
Typical power (8-Ω load)	30 W
Total Harmonic Distortion:	
Rated distortion	0.5%
IM Distortion:	
10 dB below continuous power output at 60 Hz and 7 kHz (4:1)	<0.2%
IHF Power Bandwidth:	
3 dB below rated continuous power at rated distortion	5 Hz to 50 kHz
Bandwidth at 1 W	5 Hz to 100 kHz
Sensitivity:	
At continuous power-output rating	500 mV
Hum and Noise:	
Below continuous power output:	
Input shorted	100 dB
Input open	85 dB
With 2 kΩ resistance on 20-ft. cable on input	97 dB
Input resistance	18 kΩ

Audio amplifier with IC-driven output stage

It is pointed out that the all discrete-device stereo amplifier can become quite complex because of the many direct-coupled transistor stages required. Building and placing such amplifiers in operation can tax your patience and pocketbook, because it is commonplace to experience heartbreaking catastrophes until all bugs are finally eliminated. Even at best, most designs are prone to oscillation problems; these problems often wipe out the power stage together with several of the driver and low-level circuits. Much trouble arises from physical layout, grounding, and connections because a stereo amplifier has both high-gain and high-frequency capabilities. Adding to these inherent problems is the sloppy gain tolerances on transistors. In ac-coupled circuits this wouldn't be so difficult to cope with, but in direct coupling, it often takes a bit of trial and error to get all transistors within their proper operating and biasing ranges. As most hobbyists have learned, the less experimentation needed, the less likely there will be a horror story.

A neat solution to this dilemma is to use a dedicated IC in place of the whole assemblage of driver and low-power transistors generally found in stereo circuits.

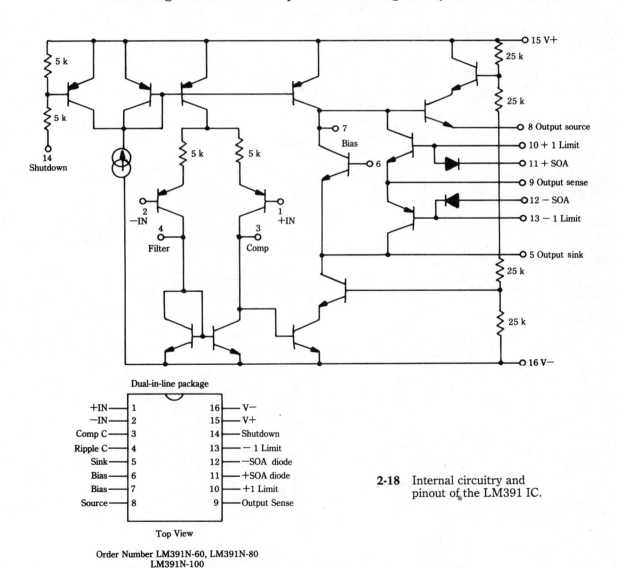

2-18 Internal circuitry and pinout of the LM391 IC.

The National Semiconductor LM391 designates a series of such ICs. By using one of these very specialized ICs, it is possible to design a stereo amplifier devoid of the usual array of low-level transistors. The internal circuitry of this IC is shown in Fig. 2-18, and Fig. 2-19 depicts a 60 W amplifier making use of only this IC and the Darlington output stage—there are no other active devices.

A further simplification could be readily made by selecting appropriate monolithic Darlington devices for the output. On a relative basis, however, much simplification has been achieved by the elimination of a half-dozen, or so, transistors. Inasmuch as these transistors are now within the LM391 IC, the stereo amplifier is electrically equivalent to traditional discrete designs, but much of it has been pre-engineered. This not only lifts a burden from the hobbyist, but reduces costs for the manufacturer.

Being a dedicated IC, the LM391 family yields a performance format that greatly exceeds that attainable from garden variety op amps. Drive capability embraces 10 to 100 W stereo designs. Input noise is approximately 3 μV and dis-

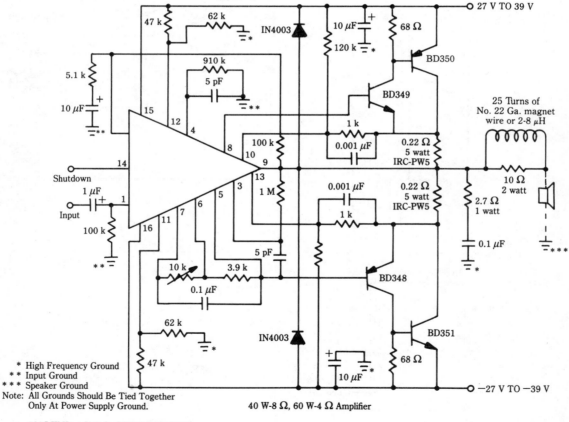

2-19   A 60 W amplifier using the LM391 IC. National Semiconductor Corp.

tortion is on the order of 0.01 percent. The IC is internally protected for output faults and thermal overloads, and circuitry for protecting the stereo output transistors is user programmable. Similarly, gain and bandwidth are readily selectable by the user. Further stereo protection is provided by dual-slope SOA protection and there is a shutdown pin. Rejection to the effects of supply-voltage variation is 90 dB. It is evident that the simplification provided by this IC will usually be accompanied by performance enhancement over traditional discrete designs. The pinout of the LM391 is shown by Table 2-2.

Table 2-2. Circuit function of pins on the LM391 audio power drive.

Pin No.	Pin Name	Comments
1	+Input	Audio input
2	−Input	Feedback input
3	Compensation	Sets the dominant pole
4	Ripple Filter	Improves negative supply rejection
5	Sink Output	Drives output devices and is emitter of AB bias V_{BE} multiplier
6	BIAS	Base of V_{BE} multiplier
7	BIAS	Collector of V_{BE} multiplier
8	Source Output	Drives output devices
9	Output Sense	Biases the IC and is used in protection circuits
10	+Current Limit	Base of positive side protection circuit transistor
11	+SOA Diode	Diode used for dual slope SOA protection
12	−SOA Diode	Diode used for dual slope SOA protection
13	−Current Limit	Base of negative side protection circuit transistor
14	Shutdown	Shuts off amplifier when current is pulled out of pin
15	V+	Positive supply
16	V−	Negative supply

National Semiconductor Corp.

120 W quasi complementary-symmetry stereo amplifier

A stereo system with 120 W per channel capability is a good compromise among a number of factors. This power level is high enough to do justice to music peaks, but sufficiently modest to hold down expenses. The overall performance of amplifiers with power ratings in this vicinity is sometimes better than is readily forthcoming from higher power systems. This is because it tends to be easier to obtain good linearity, and low hum and background noise at room volume from moderate-power amplifiers than from high-power amplifiers. The 120 W amplifier shown in

Fig. 2-20 utilizes a quasi complementary-symmetry output stage with four identical power transistors in push-pull parallel. The quasi complementary-symmetry circuit has a good track record in stereo practice; the operating balance between the two sections of the output stage is surprisingly good even though one section delivers power from the collectors, and the other section delivers power from the emitters of the output transistors. A worthwhile feature of this circuit is that the output transistors are directly coupled to the 8-ohm speaker—no large electrolytic coupling capacitor is needed. This improves low-frequency response, feedback stability, and reliability.

Note the driver stages for the paralleled output transistors in Fig. 2-20. For the upper pair of paralleled RCA1B04 power transistors, the driver is a Darlington-

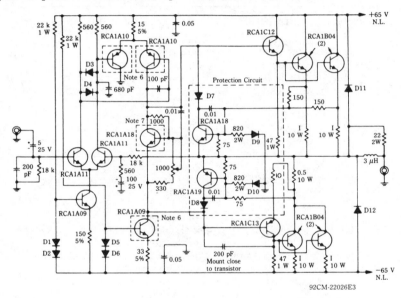

92CM-22026E3

NOTES:

1. D1–D8 – 1N5391; D9,D10 – 1N914B; D11, D12 – 1N5393
2. Resistors are 1/2 W ± 10% unless otherwise specified; values are in ohms
3. Capacitances are in μF unless otherwise specified.
4. Non-inductive resistors
5. Provide approx. 1°C/W heat sinking per output device based on mounting with mica washer

and ZnO thermal compound (Dow Corning No. 340, or equivalent) with $T_A = 45°C$ max.
6. Mount on heat sink, Wakefield No. 209-AB, or equivalent. (Alternatively, this type may be obtained with a factory-attached integral heat sink.)
7. Attach heat sink cap (Wakefield No. 260-6SH5E, or equivalent) on device and mount on same heat sink with output transistor.

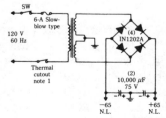

NOTES:

1. 93°C thermal cutout (attached to heat sink for output transistors (Elmwood Sensor part No. 2455-88-4), or equivalent.
2. Power transformer: Signal 88-6, Signal Transformer Co., 1 Junius St., Brooklyn, N.Y. 11212, or equivalent. Use 125-V primary tap.

2-20   A 120 W quasi complementary-symmetry audio amplifier.

connected RCA1C12, an NPN transistor. The driver for the lower pair of power transistors is an RCA1C13, a PNP transistor. Here you also have a Darlington connection, but somewhat different from the more conventional Darlington circuits where both transistors are of the same type. It is this difference that causes the lower pair of RCA1804 output transistors to operate as if they were PNP, rather than NPN types. This explains the description of the output stage as having quasi complementary-symmetry. In other words, the output stage simulates a push-pull circuit using NPN and PNP transistor types in the output stage.

Continuing the sequence of the circuit analysis from the output to the input of the amplifier, the next stage in the signal path is the differential pair comprising the RCA1A10 PNP transistors. This stage is associated with a constant-current source, the RCA1A09, which is shown enclosed in a dashed-line box. The RCA1A18 transistor has a unique function and will be subsequently dealt with. Finally, you arrive at the input stage, the differential amplifier circuit comprising the RCA1A11 NPN transistor pair. This stage also operates from a constant-current source, again an RCA1A09 transistor. The incoming audio signal feeds one base of the differential input stage; the other base of this stage is impressed with negative-feedback from the output stage.

The transistors thus far accounted for are involved in amplification of the audio signal. Additionally, there are three other transistors that serve important functions in auxiliary circuitry. One of these transistors and its associated circuitry have been referred to but not described. This is the stage involving the RCA1A18 NPN transistor. The base-emitter voltage drop in this transistor establishes the dc operating bias for RCA1C12 and RCA1C13 driver transistors. Because of the direct coupling between amplifier stages, this technique also provides the bias for the output transistors. Output transistor bias is a critical operating parameter in stereo amplifiers. If the output transistors are allowed to operate in pure class B, efficiency will be maximized but crossover distortion will be objectionably high. At the other extreme, class A operation is not permissible because of the inordinately low efficiency. The best mode is a compromise—class AB, with just enough forward bias to bring crossover distortion down to acceptable levels. Forward bias beyond this impairs efficiency and causes thermal problems.

The reason the output stage bias is critical is because it is temperature sensitive. What is needed is an adjustable bias source that temperature-tracks the output transistors. This, indeed, is the function of the RCA1A18 and associated circuitry. The two important features of this circuitry is that it operates from a constant-current source, and the RCA1A18 transistor is mounted on the heat sink of one of the output transistors. With such thermal feedback and constant current, the voltage developed across the base-emitter junction of this transistor temperature-tracks the required bias of the output transistors. The degree of class AB operation is adjustable via the 1000-ohm potentiometer in this circuit. If the idling current of the output stage is thereby adjusted, it will remain nearly constant over a wide range of temperature variation.

The second auxiliary circuit comprises the two transistors within the dashed box labeled *Protection Circuit*. The function of these transistors is to drastically reduce the gain of the amplifier if an attempt is made to force the output much over the rated power level of 120 W. A study of the circuitry will show that these transistors are active loads for the second amplifier stage (the differential stage, configured around the pair of RCA1A10 PNP transistors). It will be seen that the base circuits of these protective transistors sense the voltage drops developed across the 10-watt resistors in the output circuits of the power stages. When output current tends to reach excessive levels, the protective transistors become more conductive. This depletes the gain of the second amplifier stage, and therefore of the entire amplifier. As a result, output current can only increase slightly, rather than proportionally, as would be the case without the protective circuitry.

Servo or motor-drive amplifier with 120 V output

A common practice in servo- and motor-drive systems is to adapt an audio amplifier for the purpose. This tends to be a convenient approach because of the prevalence of low-cost, powerful stereo equipment. This isn't always as straightforward as initial consideration might suggest, however. Audio amplifiers sometimes don't stand up when called upon to deliver sustained power at their advertised ratings. This is because the average power demanded from them in handling audio signals is relatively low due to the peaked waveforms involved: Also, in room listening, the volume control is generally set much below that corresponding to high-power delivery. The broad frequency response is not needed, and may even prove troublesome. Usually a transformer must be procured that will provide the needed voltage for the motor—often 120 V ac.

In light of the above problems, it may be more desirable to build an amplifier especially designed for servo systems and motor-drive purposes. The amplifier shown in Fig. 2-21 is such a circuit. The push-pull Darlington output stage delivers

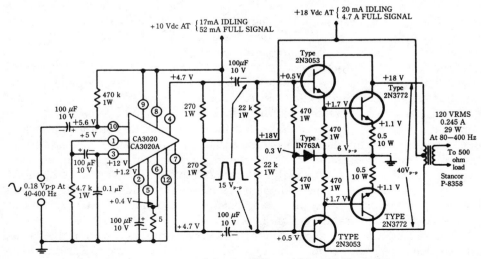

2-21 A servo- or motor-drive amplifier with a 120 V output. RCA

its power into a transformer, so that 120 V are available for the motor. A push-pull drive signal is provided by the CA3020 IC, and no interstage coupling transformer is needed. The diode in the base-return circuit of the Darlington stage breaks the path of the destructive current that might otherwise flow from avalanche breakdown in the base-emitter sections of the 2N3053 drive transistors. Note that this is not a negative-feedback amplifier because distortion is not considered a problem in most servo- and motor-drive applications. Because distortion is not a problem, operation is possible at a higher gain than would otherwise be the case. The output stage operates class B and no detrimental effect is generally experienced from crossover distortion. Fewer headaches are likely to be encountered from an amplifier of this type than from a wideband feedback amplifier of the type used in audio work.

In order to provide additional insight into this amplifier, the block diagram and schematic circuit of the CA3020 are shown in Fig. 2-22. The output transistors of this IC, Q6 and Q7, have peak current capabilities of 250 mA. This, together with the push-pull output configuration makes this IC an excellent driver for a more powerful push-pull stage.

The RCA HC2000H power op amp

Op amp ICs are usually small-signal devices; their traditional function has been to provide the precision processing of linear relationships between various parameters. If power is thereafter needed, it has been customary to follow the op amp with a power device.

The RCA HC2000H hybrid power module is a notable exception to this practice. This interesting device is a true op amp, having differential input, high gain, and precisely tailored stages. But, unlike its many low-power monolithic versions, the HC2000H can deliver 100 W RMS into a 4-ohm load, and can provide 7 A peak currents.

The simple amplifier circuit shown in Fig. 2-23 might fail to reveal the true nature of this device which possesses both the complexity and the sophistication of its better known low-power relatives. This can be grasped from an inspection of the internal circuitry shown in Fig. 2-24. The output stage operates in class B and utilizes a quasi complementary-symmetry configuration that permits direct coupling to the load when a split-polarity power supply is used.

Built-in protection is provided against a short-circuited load and against the effects of reactive loads. The metal hermetic package is both convenient and efficient for use with thermal hardware.

Looking at this device as an op amp, its open-loop gain is in the vicinity of 2000; its closed-loop gain is typically 30. (The feedback path is internal, but is brought out to a connecting wire so that both open- and closed-loop tests can be made.) In practice, the module can provide 60 W from dc to 30 kHz under closed-loop conditions.

Designers and experimenters would do well to consider this device for applications involving audio power amplifiers, power operational amplifiers, servo amplifi-

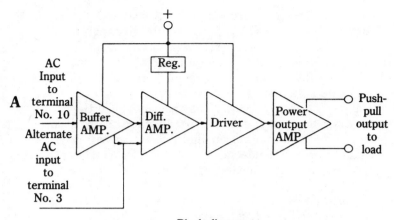

Block diagram.

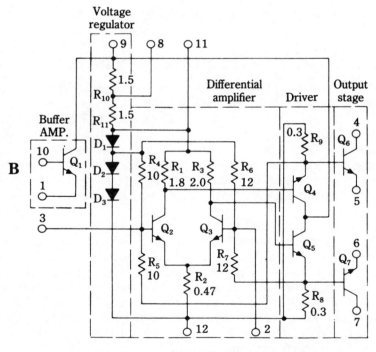

Schematic diagram.

2-22   The CA3020 integrated circuit. ʀᴄᴀ

ers, deflection amplifiers for CRTs (cathode ray tubes), solenoid drivers, linear
voltage regulators, and similar power techniques. A single power supply with a dc
output of 30 to 75 V might be used in place of the split supply; in such a case, the
load should be coupled to the output (terminal 3) through a 2000 μF electrolytic

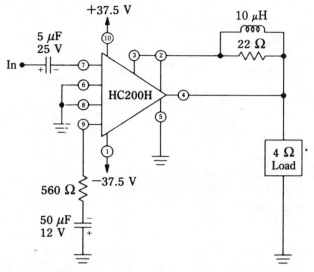

2-23 RCA hybrid power op amp with 100 W output. RCA

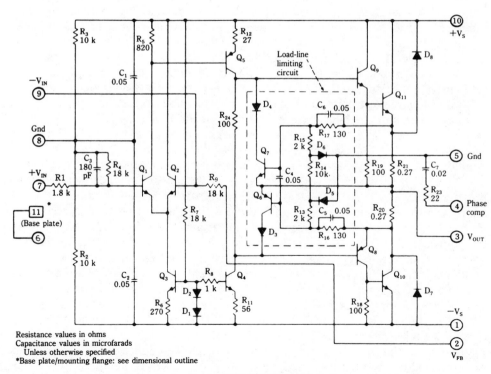

Resistance values in ohms
Capacitance values in microfarads
  Unless otherwise specified
*Base plate/mounting flange: see dimensional outline

2-24 Internal circuitry of RCA HC2000H power op amp. RCA

capacitor, with the plus lead of the capacitor connecting to terminal 3 of the HC2000H. And, of course, terminal 1, which connects to the negative voltage source when using a split supply, is grounded when using a single dc supply.

RF amplifiers for mobile service

The secret of successful power amplification at higher frequencies is to select a transistor that has been designed for the specific type of service to be used. Although a high-frequency transistor will always work at lower frequencies, certain optimizations can be lost in this way. Other things being equal, lower-frequency transistors tend to be more electrically rugged, tend to have greater power capability, and tend to cost less per watt. By using transistors intended for the frequency or band of frequencies needed, it is less likely that trouble will be encountered with self-oscillation or other instabilities. Additionally, if the designated RF transistor shown in a schematic diagram is used, the various reactances and resonant circuits depicted will more likely be on target than if another type of transistor is substituted. Many RF transistors have input-matching networks within their packages; these are obviously frequency sensitive. You cannot build high-frequency amplifiers with the reckless abandon that is allowable for low-frequency projects.

Several examples of RF power amplifiers suitable for mobile operation from 12 V systems are described. The RF transistors used in these amplifiers are not only optimized for the frequencies involved, but they are specifically designed for operation from 12 V storage-battery systems in vehicles. Thus, when striving for rated power output, one can be certain that the current is also within specified ratings. Not only does this prevent damage to the transistor, but it ensures that there will be no drastic drop in current or power gain as full output power is approached.

These amplifiers are for FM systems. Linear (class B) AM service is possible. With due regard to ratings, they can also be applicable to SSB (single sideband) operation with suitable bias circuits. Use of these amplifiers for SSB is not recommended because the transistors used have been optimized for linearity. (Linearity requirements from a class B amplifier are less demanding for AM than for SSB.) Note that all amplifiers have 50-ohm input and output ports.

All of the amplifiers have LC networks in their positive leads to discourage low-frequency oscillation. LC networks are necessary because RF transistors operate on the slope of their current-gain curve, which means that their current gain increases with decreasing frequency so that at audio and low radio frequencies, the transistor is willing to oscillate with very little provocation. From a practical standpoint, this tendency is opposite to the tendency of tube amplifiers to generate parasitic oscillations at very high frequencies, and often comes as a surprise to those whose RF experience is with tube transmitters. Note also that no neutralization is needed for any of these amplifiers. Here again it is important to use RF power transistors have relatively high inductance per unit length, whereas conductors with their special packaging techniques. A switching-type power transistor might possess the frequency capability for a certain RF application, but because of high internal feedback, would likely give trouble with self-oscillation.

A mobile 70 W, 50 MHz amplifier

For a solid-state amplifier at 50 MHz, 70 W is a respectable power level, one which not too long ago was not attainable from a single consumer-oriented device. The PT8854 transistor features an inordinately low thermal impedance—only 0.86°C/W typical. It is able to efficiently transfer dissipated power to the chassis or heat sink. Making use of this RF power transistor, the amplifier shown in Fig. 2-25 develops 10 dB of power gain and requires 7 W of RF drive. In common-emitter RF amplifiers and especially in higher-power implementations, the slightest amount of lead inductance connecting the emitter to ground will produce degeneration or negative feedback, tending to reduce power gain. This should receive top priority in establishing the layout pattern for the amplifier components. Another reason for keeping emitter inductance low is that the feedback can turn positive at some frequency, leading to instability. Keep in mind that small-diameter conductors have relatively high inductance per unit length, whereas conductors with large diameters or large cross sections tend to have lower inductances. Wide straps are very good and are often used as conductors. In any event, the physical length of the path from emitter connection(s) to RF ground should be very short.

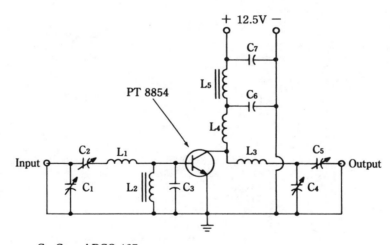

C_1, C_2	ARCO 467
C_3	1000 pF Underwood Electric
C_4, C_5	ARCO 465
C_6, C_7	.1 μF

L_1	1 Turn No. 18, 5/16 inch diam.
L_2	2 1/2 Turns No. 24 on Ferroxcube VK211-07-3B Core
L_3	2 Turns No. 16 5/16 inch diam.
L_4	3 Turns No. 16 5/16 inch diam.
L_5	2 1/2 Turns No. 16 on Stackpole Ceramag No. 9500, D0 A7 Core

2-25 A 75 W, 50 MHz amplifier for mobile service. TRW

A mobile 25 W, 90 MHz amplifier

The RF amplifier shown in Fig. 2-26 essentially repeats the basic theme of the previous amplifier. The fact that competitive semiconductor firms are involved reveals common techniques of power boosting at high frequencies. This apparent standardization stems not so much by mutual consent, as by the harsh dictates of technology and economics. This wasn't always so; at first, power amplification at RF was more of a black art in which the experimenter or designer tried a hand at coaxing power, stability, and reproducibility from existing transistors with high f_T ratings. Usually, one objective had to be severely compromised to achieve acceptance in the others. In contrast, you can now order a dedicated RF power transistor and build an amplifier according to a rigorously designed circuit which will produce reasonably predictable results. Minimal experimentation and tweaking is necessary if construction is carried out with observance of recognized high-frequency practices.

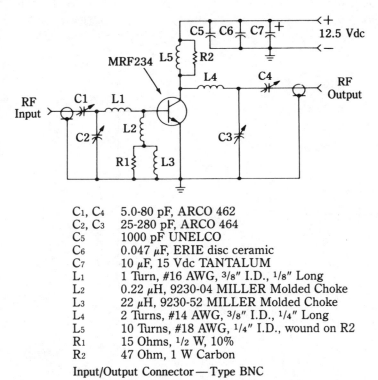

C_1, C_4	5.0-80 pF, ARCO 462
C_2, C_3	25-280 pF, ARCO 464
C_5	1000 pF UNELCO
C_6	0.047 μF, ERIE disc ceramic
C_7	10 μF, 15 Vdc TANTALUM
L_1	1 Turn, #16 AWG, 3/8" I.D., 1/8" Long
L_2	0.22 μH, 9230-04 MILLER Molded Choke
L_3	22 μH, 9230-52 MILLER Molded Choke
L_4	2 Turns, #14 AWG, 3/8" I.D., 1/4" Long
L_5	10 Turns, #18 AWG, 1/4" I.D., wound on R2
R_1	15 Ohms, 1/2 W, 10%
R_2	47 Ohm, 1 W Carbon

Input/Output Connector — Type BNC

2-26   A 25 W, 90 MHz amplifier for mobile service. Motorola Semiconductor Products, Inc.

A mobile 10 W, 175 MHz amplifier

VHF (very high frequency) amplifiers might still be constructed with lumped-circuit reactances and resonant tanks. However, exceptional care must be directed towards lead lengths, spacings, and physical dimensions. At these frequencies, stray capacitance and parasitic inductance exert considerable influence on resonating frequencies, impedance matching and transformation, bypassing, and coupling. It is an in-between frequency region where stripline techniques are tantalizingly desirable, but often physically awkward. Lumped-circuit elements are satisfactory if they are physically compact and exhibit good electrical and mechanical integrity. The named brands of the components listed for this circuit represent specialists in high-frequency components; it would probably be unwise to make substitutions.

Because of antenna gain factors at both transmitting and receiving stations, a 10 W amplifier at 175 MHz, such as the one shown in Fig. 2-27, is often capable of very reliable communications service. In spite of antenna gain, the short distances intended for coverage by many mobile services often do not justify more than 10 W output. This circuit is relatively easy to drive, requiring approximately 750 mW of excitation.

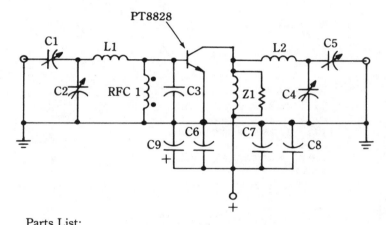

Parts List:

$C_{1, 2, 4, 5}$	Trimmer car, ARCO #462, 5-80 pF
C_3	120 pF Underwood Mfg.
C_6	1000 pF Underwood Mfg.
C_7	0.01 μF disc ceramic.
C_8	0.02 μF disc ceramic.
C_9	25 μF, electrolytic, 35 WVDC.
L_1	2 T., #18 AWG., 0.25" I.D.
L_2	2 T., #18 AQG., 0.25" I.D.
Z_1	8 T., #18 AWG., wound on 330 ohms $1/2$ W resistor.
RFC1	$2^1/_2$ T., #22 AWG. on Ferrox cube VK211-17/4B Core.

2-27 A 10 W, 175 MHz amplifier for mobile service. TRW

A mobile 20 W, 470 MHz amplifier

The amplifier shown in Fig. 2-28 differs from the previous circuits in that stripline elements are used in place of lumped circuits for resonating and impedance match-

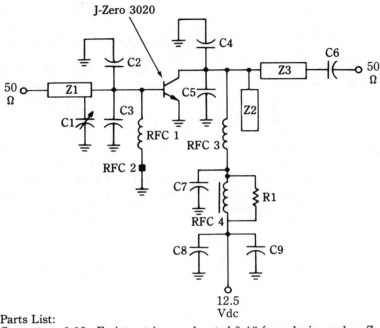

Parts List:

C_1	1-10 pF piston trimmer, located 0.4″ from device end on Z_1.
$C_{2,4}$	30 pF uncased mica capacitor.
C_3	20 pF uncased mica capacitor.
C_5	40 pF uncased mica capacitor.
C_6	30 pF miniature uncased mica capacitor.
C_7	220 pF ncased mica capacitor.
C_8	5 μF @ 25 volts.
C_9	0-1 μF @ 50 volts disc ceramic.
RFC 1	6 T. .18 AWG., 1/4″ diam., 1/2″ long.
RFC 2	Ferrite bead.
RFC 3	4 T. .18 AWG., 3/8″ dia.; 1/2″ long.
RFC 4	3 T. .18 AWG., on 1/2″ "H" toroid
R_1	10 ohms @ 1/2 watt.
Z_1	Microstripline, $Z_0 = 30$ ohms, W = 3.37″, L = 3.02″
Z_2	Microstripline capacitor, $Z_0 = 24$ ohms, W = 0.46″, L = 1.62
Z_3	Microstripline, $Z_0 = 20$ ohms, W = 0.56″, L = 2.34″

Board material = 1/16″ TFE-glass

2-28 A 20 W, 470 MHz amplifier for mobile service. TRW

ing. Note also that RF choke action is obtainable from the length of a conductor passed through a ferrite bead. Lower RF losses and better mechanical and thermal stability are more readily forthcoming from stripline reactances and resonators than from lumped circuits in this frequency range, and the power capabilities of dedicated UHF (ultra high frequency) transistors can be effectively exploited. Actually, this amplifier can, by minor physical or electrical modification, be optimized for performance throughout the 450–512 MHz range. You might suppose that the output and frequency capability of this transistor is marginally attained;

this is not the case, as attested by the rated collector efficiency of 60 percent—this ranks with the best in RF power transistors. About 3.5 W of input power are needed.

When constructing amplifiers with stripline elements, it is of utmost importance to use the exact board material and thickness that is specified; otherwise, resonances and reactances will not duplicate those of the original circuit. Additionally, it is possible that RF losses will be unacceptably increased.

Hybrid power-boost module for amateur two-meter band

Amateur two-meter transmitters and transceivers often operate at power levels in the fractional-watt to several-watt range. A booster amplifier with a 25 W capability is useful for improving communications reliability. The construction of such an amplifier can be greatly simplified, and considerable time and experimentation can be avoided by using a dedicated hybrid module intended for this purpose. This is the Motorola MHW252 RF power amplifier module. It is broadbanded to accommodate the 144 to 148 MHz range and has a minimum specified power gain of 19.2 dB. However, typical units develop sufficient gain to boost a 250 mW input to 25 W output.

The MHW252 comprises two cascaded power stages and features thin-film hybrid construction with nichrome resistive elements, beryllium-oxide substrates, and NPO (negative-positive-zero), MOS (metal-oxide semiconductor), and GP dielectric capacitors. All of the devices within the module are glass passivated. Computer designed impedance-matching networks are used, and 50-ohm input and output impedances are maintained across the band. The internal transistors are produced on separate production lines where their parameters are optimized for the specific requirements of such a power-boost module. Thus, the manufacturer is able to specify stable performance with any load VSWR up to 6:1, at any phase angle. At the same time, output harmonics of the module itself (no output filter) are 30 dB below the output level.

From the standpoints of performance, convenience, and cost, it would be difficult to start from scratch with discrete components and equal the results that are readily available from this pre-engineered module. Although intended for the FM modulation format that has long been used on two meters, the enterprising experimenter might be able to obtain satisfactory results in other modes. The schematic diagram of the power booster is shown in Fig. 2-29. Note that the booster is associated with an input attenuator, an output filter, and a carrier-operated relay.

The input attenuator enables 25 W of power to be developed for various input power levels. By its proper deployment, it protects the module against input overload when the available drive power exceeds 200 mW or so. The input attenuator can be implemented to accommodate a maximum of 5 W of input power. In all cases, this attenuator presents 50 ohms to the drive source. Table 2-3 lists the values of the resistors comprising the input attenuator. R1 and R2 should be 2 W carbon types; R3 can be a 1 W carbon resistor.

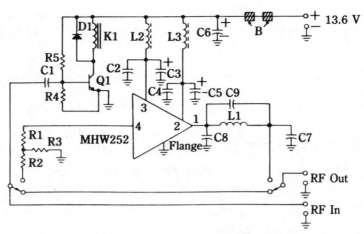

R1, R2, R3—See Table 2-3
R4—1.0 k/¼ W
R5—27 k/¼ W
C1—5.0 pF Dipped Mica
C2, C4—0.01 μF Ceramic Disc
C3, C5—1.0 μF Dipped Tantalum
C6—10 μF/15 V Electrolytic
C7—24 pF Dipped Mica
C8, C9—10 pF Dipped Mica

L1—1¼Turns AWG #18 Enameled
  Wire ⁵/₁₆″ ID
L2, L3 Ferroxcube VK200—20/4B or
  Equivalent
K1—Arrow MNFZ - 12, Omron
  LZN2-UA-DC12 or Equivalent
Q1—2N4401
D1—1N4001
B—Ferrite Beads, Ferroxcube 5659065/3B
  or Fair-Rite Products Corp. 2673000101
  or Equivalent.

2-29   A 25 W power booster for the amateur two-meter band. Motorola Semiconductor Products, Inc.

Table 2-3. Attenuator resistances for different input power levels.

Power input (W)	Power ratio	Attenuation (dB)	R1, R2(Ω)	R3(Ω)
0.20	1.0	0	Short	Open
0.50	2.5	4	11	100
0.80	4.0	6	16	68
1.00	5.0	7	20	56
1.50	7.5	9	24	39
2.00	10.0	10	27	36
2.50	12.5	11	30	30
3.00	15.0	12	30	27
4.00	20.0	13	33	24
5.00	25.0	14	36	22

The input power applied to the module must not exceed 300 mW.

The output filter is comprised of inductor L1 and capacitors C7, C8, and C9. Although this low-pass filter has a Q of less than unity, it attenuates the second harmonic by 60 dB and the third harmonic by 50 dB. It should be noted that this filter has an M-derived configuration, that is, one of its elements is resonated. Specifically, L1 is tuned by C9 to resonate at approximately 292 MHz. Thus, in addition to the filter normal attenuation of second- and third-harmonic energy, extra attenuation is accorded to second-harmonic frequencies of the two-meter band. The leads of C7, C8, and C9 should be short in order to minimize the parasitic inductance of these capacitors; otherwise, the described results might not be obtained.

Carrier-operated relay K1 is energized when RF applied to the input of Q1 is rectified in its base-emitter diode. Note the tiny capacitor, C1, used to couple the RF signal to Q1. Q1 is forward biased to about 0.6 V in order to increase its sensitivity. Diode D1 suppresses voltage transients which could endanger Q1.

The power module has an internal dc decoupling network which suppresses any tendency for low-frequency oscillation, down to 5 MHz. External decoupling circuitry is needed to suppress lower-frequency oscillations. Recall that RF power transistors, such as those used in the module, develop much higher gain at low frequencies than at their operating frequency. Precautions must therefore be taken against low-frequency parasitics—such oscillation can be destructive to the module. Inasmuch as this module has two stages of amplification, terminals for positive dc supply are provided for each of them. Each stage must have its individual dc decoupling network, both internally and externally. In the schematic diagram of Fig. 2-29, one of the external dc decoupling networks is comprised of L2, C2, and C3. The other is made up of L3, C4, and C5. Additionally, a mutual dc decoupling network consists of ferrite beads B in conjunction with capacitor C6. This seems to be a complicated way to provide dc power for the module. It is, however, an important procedure for proper and safe operation.

In RF work, board layout and fabrication techniques are not incidental—at VHF and higher frequencies, the orientation and spacing of parts is as important as following a schematic diagram. The PC board and parts location are shown in Fig. 2-30. For best results, these should be duplicated or at least followed in a general way.

A 300 W, 2 – 30 MHz MOSFET linear amplifier

A PEP (peak envelope power) of 300 W is considered a practical power level for reliable single-sideband communications in the amateur radio HF (high frequency) bands. A linear amplifier capable of providing this power with no tuning controls throughout the 2–30 MHz range has become an expected luxury. The MOSFET push-pull linear amplifier shown in Fig. 2-31 meets this criteria. Operating from a 100 V dc source, the amplifier exhibits tubelike qualities in its circuit requirements. For example, the impedance levels are very manageable—they are not inordinately low as is often the case in bipolar transistor circuits. This makes it easy to implement tank circuits with practical valued elements. In this amplifier there are no resonant networks; only baluns and transformers are used in the input and output circuits.

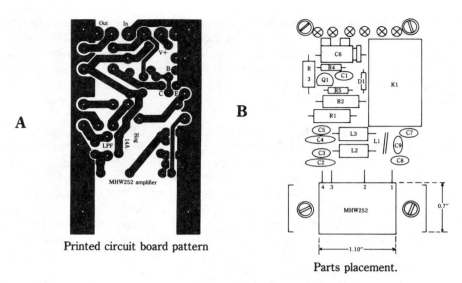

A

Printed circuit board pattern

B

Parts placement.

2-30 Constructional guidance for two-meter, 25 W power booster. Motorola Semiconductor Products, Inc.

There are other MOSFET advantages over bipolar transistor amplifiers. There is no need for the complex dc decoupling network often needed in bipolar designs. This is because the MOSFET displays no tendency to oscillate at low frequencies, its gain being substantially constant even over a broadbanded range. Also, there is no need for thermal-tracking bias circuitry. The MOSFET does not have the thermal-runaway characteristic of the bipolar transistor.

The two separate bias connections shown in the schematic diagram are intended to facilitate adjustment of equal idling currents for the two MOSFET devices. A single bias supply in conjunction with a potentiometric arrangement can be used (Fig. 2-32). Initially, R1 should be adjusted to produce about 4 V at point A. Then, the two remaining pots can be used to provide balanced quiescent current in the two MOSFET devices.

In any event, in contrast to bipolar-transistor bias, the MOSFETs do not consume dc bias current. The idea of bias is to project the operation of the amplifier a bit into the AB region because pure class B operation is attended by stronger harmonic generation and intermodulation products. Between 2 and 3 V of positive bias should suffice to produce the suggested 0.25 A of idling current per device.

Note that the same type of core is specified for all four major core components. The inset shows the stacking arrangement of the cores for T1, T2, T3, and T4. When placed in use, a different pi-section low-pass filter must be inserted between the amplifier and the antenna for each band. This is standard practice and is needed to further attenuate harmonic energy and intermodulation products in order that these will not be radiated and cause interference. Figure 2-33 depicts element values for suitable filters. (L1 and L2 can be empirically found by resonating with a capacitor equal to C1/2.)

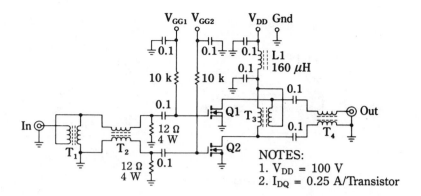

T₁ 4:1 Transformer
    4 Turns Of Two 50 Ω Coax In Parallel Through
    Six Indiana General Ferrite Cores PN F627-8-Q1

T₂ 12.5 Ω Balun
    2 Turns Of Four 50 Ω Coax In Parallel Through
    Six Indiana General Ferrite Cores PN F627-8-Q1

T₃ 4:1 Transformer
    4 Turns Of Two 50 Ω Coax In Parallel Through
    Six Indiana General Ferrite Cores PN F627-8-Q1

T₄ 50 Ω Balun
    6 Turns Of 50 Ω Coax Through Six Indiana
    General Ferrite Cores PN F627-8-Q1

Core Arrangement
T1, T2, T3, T4

CAUTION: Beryllium Oxide—The top cap of this device is alumina which is harmless. However, the ceramic portion between the leads and the metal flange is Beryllium Oxide, the dust of which is toxic. Care must therefore be taken during handling and mounting the device to prevent any damage to this area.

Steps must be taken to ensure that all those who may handle, use, or dispose of this device are aware of its nature and of these necessary safety precautions. In particular the transistor should never be thrown out with general industrial or domestic waste.

2-31 A 300 W 2–30 MHz linear amplifier with push-pull power MOSFETs. Siliconix, Inc.

The power gain over the 2–30 MHz range is a nominal $17\frac{1}{2}$ dB $\pm$ $\frac{1}{2}$ dB. Accordingly, the amplifier can be driven from a 50-ohm source with 6 W PEP capability.

A BIMOS power switch

At certain combinations of voltage and frequency, designers of industrial equipment have long faced a cost and performance dilemma. When it has been advantageous to operate at switching speeds between, say 20 and 200 kHz and control several to several tens of kilowatts in 500 to 1000 V loads, what is the appropriate power device?

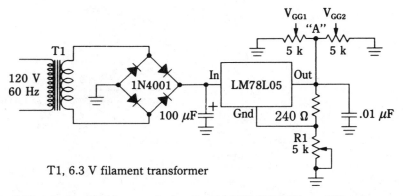

2-32   A suitable bias supply for the 300 W MOSFET amplifier. Siliconix, Inc.

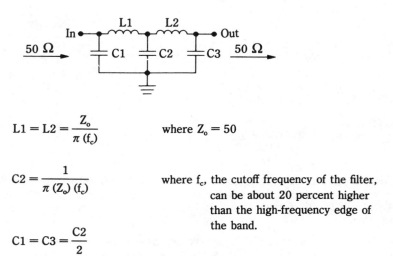

$$L1 = L2 = \frac{Z_o}{\pi \, (f_c)}$$ where $Z_o = 50$

$$C2 = \frac{1}{\pi \, (Z_o) \, (f_c)}$$ where f_c, the cutoff frequency of the filter, can be about 20 percent higher than the high-frequency edge of the band.

$$C1 = C3 = \frac{C2}{2}$$

2-33   Data for design of output filters for 300 W linear amplifier.

The answer directly affects a large market comprising high-frequency bridge inverters for motor drives, uninterruptible power supplies, 440–480 V off-line equipment, off-line switching power supplies, and welders. Equipment operating from three-phase power lines also tends to fall into this general classification.

Bipolar power transistors capable of good switching performance at 25 kHz or so, are very expensive when you specify a safe operating area in the neighborhood of 800 V or higher. A typical bipolar power transistor with 10 A capability is found to have a V_{CEO} rating of only 400 V when used in conventional power switching circuits (in the common-emitter connection). The trade off for higher V_{CEO} capability is cost. The exotic SOA ratings above 400 or 500 V are accompanied by exotic prices.

It would be natural to suppose that the power MOSFET is the answer, inasmuch as this newer device outperforms bipolar power transistors in so many applications. The drawback here is that these devices develop wasteful voltage drops when rated for high-voltage service. For example, an 850 V power MOSFET would drop about 18 V, and a MOSFET with 1000 V capability would show a nominal 23 V drop from drain to source. (These are 10 A load situations.) By comparison, the bipolar transistor has a $V_{CE(sat)}$ of only 2.5 V under the same load conditions. This is a tantalizing situation, for what is needed is a device with the switching and high-voltage capabilities of the power MOSFET, but with the relatively low voltage drop of the bipolar power transistor. (Recall that power dissipation increases as the square of the voltage drop across the switching device.)

There are, to be sure, other devices which merit consideration. The GTO (gate turn off) thyristor can handle large power at a low voltage drop and dissipation. This device shows considerable promise, but best applications are in low-frequency circuits. Also, many designers do not like to contend with the commutation problems that sometimes accompany all thyristor circuits. That is also why inverter-rated SCRs tend to take second place to bipolar transistors, Darlingtons, and MOSFETs whenever these devices can be pressed into service as power switches.

Fortunately, a power switch is now available for reliable and cost effective use in the voltage and frequency combinations where other devices tend to produce awkward compromises. It is a composite circuit called a BIMOS (bipolar metal-oxide semiconductor). The basic arrangement is shown in Fig. 2-34.

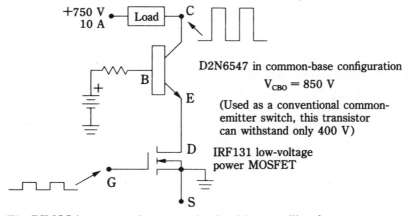

2-34 The BIMOS is a composite power circuit with compelling features. International Rectifier Corp.

In this arrangement, the power switch exhibits the high-voltage and high-frequency capabilities of the bipolar transistor together with the ease of drive of the power MOSFET. Because relatively inexpensive devices are used, the overall cost is considerably less than single devices of either type with equivalent speed and voltage ratings. Note: BIMOS is a term used by International Rectifier Corp. This company has done original development work in combining the practical features of power MOSFETs and bipolar power transistors. Their line of power MOSFETs is trademarked HEXFETs, but in this book, the generic term—*MOSFET*—is used.

In the simple circuit of Fig. 2-34, the 10 A voltage drop across the bipolar transistor is 2.5 V; with the same load, the power MOSFET develops a drop of 4.3 V. Accordingly, the total voltage drop of the composite circuit is 6.8 V. So power dissipation is low and is shared by two devices. Frequency capability is about what you would get from just a common-base bipolar circuit, but ease of drive is very nearly that of a MOSFET by itself. The circuit of the BIMOS power switch is known as a cascode configuration and has seen various services in both tube and transistor applications. The performance and cost factors displayed by the power-switch application is a relatively recent discovery. Monolithic integration of the two devices unfortunately poses some manufacturing problems. A single-package hybrid version can be expected to become available, however.

Although the circuit of Fig. 2-34 is intended to be used with square or PWM waves, a bit of experimentation might conceivably yield good results as part of a class B, or class AB audio amplifier, or as a linear or class C RF amplifier. With regard to the latter two suggestions, it is relevant to note that the common-base configuration of the bipolar transistor exhibits greater frequency capability than the more conventionally used common-emitter connection.

3
Regulated power supplies

IN COMMON WITH TRADITIONAL REGULATED SUPPLIES, MODERN CIRCUITS are of two basic types: linear and switching. *Linear regulation* is accomplished by means of variable dissipation. (There need not be anything actually linear in the operation of such regulators and the term is somewhat confusing. It is, however, commonly accepted in the technical literature to signify analog rather than digital circuitry.) *Switching regulators* operate by varying the duty cycle of a chopped wave. Newer active devices have simplified and improved the performance of both types of regulators. However, the following facts pertaining to them remain valid:

- Switching regulators are more efficient than linear types. Here, the implication of the adjective dissipative is revealing—dissipation and high efficiency are contradictory.

- Switching regulators can be packaged more compactly, particularly at higher power levels where heat sinks and blowers take up appreciable space in linear regulators.

- Above moderate power levels of approximately 100 W, switchers become more cost effective to produce and operate than linear supplies.

- For close regulation and low noise, the linear regulator is inherently superior.

The modern trend in linear regulation has been to produce all, or the greater part of the circuitry in the form of a monolithic IC. Thus there are brainy ICs such as the popular 723. This IC is a complete regulator in its own right, but is limited to about 100 mA. Used in conjunction with power devices, the overall current capability of such a simplified regulator is limited only by the rating of the external power device. Also available is a wide variety of so-called three-terminal regulators. These are power ICs generally capable of handling 1 to 10 A loads. Here again, external power devices can be added to greatly multiply current capability.

The modern trend in switching regulation has been to increase switching frequency. Long-used designs have operated in the vicinity of 20 kHz. Now with new and improved power devices, the switching frequency has been projected to the 50 to 200 kHz region. The advantages are smaller and lighter magnetics and filter components, easier containment of EMI and RFI, and lower production costs. A practical advantage sometimes is the avoidance of electrolytic capacitors because much lower capacitance is needed at these higher frequencies than at 20 kHz. With traditional switchers the bipolar transistor became very dissipative at higher frequencies; however, for a price, bipolars capable of 100 kHz operation are available. Now the trend is to use the power MOSFET in high-frequency switching supplies. Yet another important development for switchers has been availability of dedicated ICs for generating the clock frequency and processing the pulse-width modulation waveforms.

Simple, useful single-transistor power-supply circuits

Power transistors and power Darlingtons can perform useful power-supply functions in very simple circuits. Figure 3-1 depicts four configurations that have almost universal application. Because of the variable requirements of individual projects, parts values are not given. However, sufficient guidance will be provided to enable the knowledgeable experimenter to implement these circuits. All are quite tolerant to operating conditions. All can be implemented around either NPN or PNP power transistors. If the opposite-polarity transistor is used from that indicated in Fig. 3-1, just reverse dc polarities, zener-diode polarities, and electrolytic-capacitor polarities. Also, these circuits will tend to work even better with power Darlingtons if allowance is made for the fact that the voltage drop and power dissipation of Darlingtons will often be higher than that of single transistors.

The circuit of Fig. 3-1A is that of a series-pass voltage regulator. It is basically an emitter-follower with the base bias voltage stabilized by a zener diode. The output voltage is the zener-diode voltage minus the base-emitter voltage of the transistor. For many applications, 1 W zener diode with breakdown voltage in the 6 to 10 V range works out well. R can then be chosen to allow 10 to 50 mA of zener-diode current for a start. Exact design depends upon the dc input voltage, the current gain of the transistor, and the load current.

The circuit shown in Fig. 3-1B is also a voltage regulator, but here the power transistor acts as a variable shunt resistance. Voltage regulation occurs as the transistor varies its internal resistance with respect to series resistance R_S. This type of voltage regulator is very efficient for a constant load which is heavy enough to almost deplete current flow through the transistor. For lighter loads, the power dissipation in the transistor is high and the efficiency is low. Note that this characteristic of the shunt voltage regulator is just opposite to that of the previous regulator circuit which becomes progressively less efficient as loading is increased. Series resistance R_S is selected to enable regulation to prevail for about 110 percent or 115 percent of the expected load current. This is intended as a margin to allow for

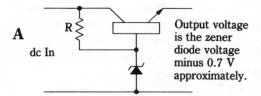

A

dc In

R

Output voltage
is the zener
diode voltage
minus 0.7 V
approximately.

Series-pass or emitter-follower voltage regulator.

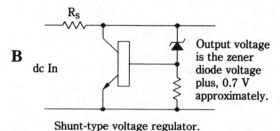

B

dc In

R_S

Output voltage
is the zener
diode voltage
plus, 0.7 V
approximately.

Shunt-type voltage regulator.

3-1 Simple single-transistor
circuits for power supply
applications.

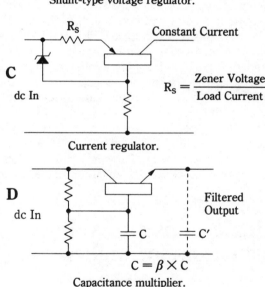

C

dc In

R_S

Constant Current

$$R_S = \frac{\text{Zener Voltage}}{\text{Load Current}}$$

Current regulator.

D

dc In

Filtered
Output

C

C′

$$C = \beta \times C$$

Capacitance multiplier.

vagaries in the dc input voltage from the rectifier filter system. In any event, such an empirical procedure will provide a good start for further optimization. The zener-diode circuit can be treated in the same way as in the previous regulator.

Via another permutation of the several components involved, the current regulator of Fig. 3-1C is obtained. By appropriate selection of the zener voltage and the value R_S, a constant current will flow through a wide range of loads. This constant current will prevail for loads from short-circuit (zero resistance) up to some maximum resistance. The product of load resistance and load current cannot exceed the dc input voltage. Indeed, this product has to be somewhat less than the dc input voltage because of the voltage drops in R_S and in the transistor itself. Other things being equal, the higher the dc input voltage, the higher the load resistance that can be accommodated. Constant-current regulators of this type find applications with electromagnetic devices and motors—for electroplating and battery charging, for electronic purposes such as the simulation of high impedance, or the linear charging of capacitors. The constant-current mode is an excellent way to operate filaments in various devices, such as transmitting tubes and projection lamps. This is because the thermionic emissivity, the optical output, and the life span of filaments are more directly related to current than to voltage.

The interesting circuit shown in Fig. 3-1D is somewhat similar to that of Fig. 3-1A. However, a capacitor has been substituted for the zener-diode voltage reference. As might be expected, this circuit no longer has the ability to behave as a dc voltage regulator. However, it now exhibits very useful ac properties. An amplified capacitance, C', now appears across the output terminals and tends to be more cost effective in filtering ripple than an equivalent physical capacitor of that size. Actually, the capacitance of the small physical capacitor, C, is multiplied by the current gain, β, of the transistor. Power Darlingtons, because of their high β values, are particularly effective in this circuit. The low output impedance to ac provided by this circuit makes it useful in stereo and TV applications where common-impedance coupling between various stages and sections must be avoided. The base-biasing resistances are chosen to produce the lowest voltage drop across the transistor consistent with the desired capacitance multiplying action. A thousand ohms for each of these resistances is a good starting point for experimentation.

It should not be supposed that the multiplied capacitance, C', has dc energy-storage properties. It is real only in its effect on ac. Manufacturers of consumer products sometimes realize worthwhile savings in both cost and bulk by using this circuit as an electronic filter in conjunction with a poorly filtered half-wave rectifier circuit.

Single-transistor voltage regulators for higher voltages

The circuits shown in Fig. 3-2 are shunt-type voltage regulators which operate in similar fashion to the shunt-type voltage regulator of Fig. 3-1B. However, in these circuits, a low-voltage zener diode is able to regulate a higher voltage. In principle, any voltage can be thus regulated if the voltage-reduction ratio provided by divider network R1, R2 is appropriate. Of course, it is also true that the shunt transistor must be selected with an adequate voltage rating. The NPN and PNP circuits of Fig. 3-2A and Fig. 3-2B, respectively,. work in essentially the same way. In previous years, not many PNP silicon power transistors were available, but now one has about the same choice of voltage and current capabilities with either type.

In implementing these circuits, be sure the current drawn by voltage-divider network R1, R2 is at least five times the zener-diode current. R_S should be as high as is consistent with the input voltage and load current involved. By making R_S

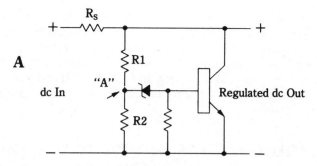

A

dc In

"A"

Regulated dc Out

High-voltage shunt regulator using an NPN transistor.

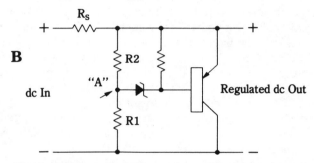

B

dc In

"A"

Regulated dc Out

High-voltage shunt regulator using a PNP transistor.

3-2 Additional single-transistor circuits.

high, minimal dissipation takes place in the transistor, thus optimizing the efficiency of the regulator. This statement, however, assumes that a fixed load is used. If there will be variation in load current, R_S must be made lower in order to accommodate the heavier loads. As previously pointed out, shunt regulators are most efficient when tailored to regulate for a fixed load.

These circuits operate in such a manner as to maintain circuit junction *A* at a potential that is the sum of the zener-diode voltage and the base-emitter voltage of the transistor. Thus, the output voltage is:

$$[(R2/(R1 + R2)](V_Z) + V_{BE}$$

Because the best zener diodes have breakdown voltages approximately in the 6 to 10 V region, these circuits provide a convenient means of obtaining reasonably good voltage regulation in the several-hundred volt region. Much, of course, depends upon the integrity of the divider resistors. Generally these should be wirewound types, or precision resistors with low temperature coefficients. In many applications, circuit B will prove advantageous over circuit A in that the transistor need not be electrically insulated from the heat sink. Whether this merits consideration depends upon the power levels involved.

There are two good features about shunt-type regulators that are often over-looked. First, these regulators draw constant current from the dc input source, regardless of load current. This tends to simplify design and implementation, as well as reduce the cost of the unregulated supply. Second, the shunt-type regulator is short-circuit proof insofar as concerns the safety of the regulating transistor. Various versions of these circuits are found in TV sets, where they are often used with Darlington transistors.

A two-terminal, constant-current diode

Constant current is required for various purposes. A common way of achieving this is to make a simple conversion of a voltage-regulated power supply. Instead of sensing output voltage, the error amplifier of the regulator is made to sense the voltage drop across a small resistance placed in series with the load. This arrangement can indeed produce very close current regulation, or can provide a preset constant current. However, it often would be more convenient to deal with a two-terminal current regulator, rather than the three-terminal (minimum) format that follows from converting the voltage regulator to a current regulator.

A number of two-terminal constant-current diodes have been on the market for quite some time; these provide constant currents in the microampere and milliampere ranges, usually being limited to currents below several tens of milliamperes. Actually, these devices are not diodes, but use either one or two JFETs (junction field-effect transistors), in the simple circuits. (Older versions used bipolar transistors.)

Naturally, it would be nice to be able to provide constant currents in the ampere range using such a two-terminal analog of the zener diode. This can be done by using a power MOSFET in conjunction with a garden-variety op amp. The circuit shown in Fig. 3-3 can provide currents from about 10 mA to 12.5 A by adjusting variable resistance *R1*. Note that unlike a conventional current-regulated power supply, there is no ground or return connection—rather, the two terminals of the circuit are simply inserted in the hot side of the load. Thus, the entire circuit behaves as a smart resistor; that is, one that automatically keeps current through itself at a constant value. This automatic-regulating feature prevails right down to a short-circuit load.

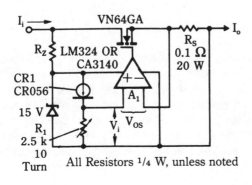

3-3 Two-terminal constant-current source using a power MOSFET.
Siliconix, Inc.

The current source supplying I_i in Fig. 3-3 must have a voltage of at least 15 V with respect to ground. The cold side of the load connects to ground. Therefore, the load receives its current through the source-drain section of the power MOSFET. Interestingly, this circuit also utilizes one of the low-current diodes referred to above. This diode, in conjunction with variable resistor R1, forms a variable-voltage, simulated-zener reference source. Coincidentally, the + and − designations on the op amp not only indicate the polarity of the dc operating voltage, but also the noninverting and inverting signal-input terminals, respectively.

Quick and easy polarity inversion

Sometimes a low-power dc source with reversed polarity from that of the main supply is needed. One way of producing such a voltage is to construct a small inverter using various transformer circuits. The output of such an inverter can then be rectified and filtered to provide the desired voltage. Tight regulation of such sources is not usually called for. They are used for such purposes as biasing substrates of certain ICs, biasing switching transistors to help sweep out stored charge during off periods (thereby increasing switching speed), and they are sometimes used for logic devices.

There is another technique for generating a reverse-polarity voltage that does not require any magnetic-core components and that is very simple and inexpensive to implement. This involves a so-called *charge pump*, which is a circuit for alternately charging and extracting charge from a capacitor. By means of a steering diode, the extracted charge polarity can be selected to be opposite to that of the charging source (the main supply). The circuit of a practical charge pump is shown in Fig. 3-4. Capacitor C is alternately charged and discharged by means of the MOSFET switch, which is driven at a 100 kHz rate by the self-excited pulse oscillator configured around the 74C14 inverter. The output of the charge pump is maintained at 5 V by the shunt-connected zener diode.

Optimum operation does not occur with a 50-percent duty-cycle switching wave from the pulse generator and a little experimentation might be in order to

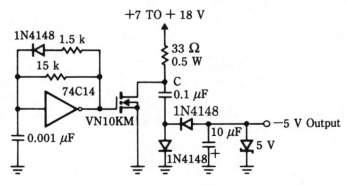

3-4 A way of obtaining a negative voltage from a positive source. Siliconix, Inc.

properly supply a given load. The 1500-ohm resistor and the 15,000-ohm resistor in the feedback path of the inverter can be slightly modified in order to achieve optimum operation. This can be done by inserting a milliammeter in series with the zener diode—the idea is to find the duty cycle of the switching wave which delivers the highest current into the zener diode. In varying the duty cycle, keep in mind that, other things being equal, the higher the switching rate the better the results. Both the 74C14 inverter and the VN10KM MOSFET perform well near 100 kHz. For equivalent results at lower frequencies, the values of capacitor C and the filter capacitor would have to be increased.

Don't neglect to connect the 74C14 to a dc operating voltage. (Such connections are generally not shown in logic-circuit diagrams.) Note that the conventional current-limiting resistance used with zener diodes is not needed in this circuit—the loading effect of the –5 V output circuit suffices for this function.

Tracking regulator using three active devices

The three-terminal IC regulator has understandably made a strong impact on power-supply technology for the simple reason that these power ICs are self-contained regulated power supplies. It is almost a matter of regarding such a subsystem as just another circuit component. Actually, successful implementation often requires some street wisdom, as discussed in chapter 1. By observing the hints, kinks, and precautions covered, it becomes essentially a matter of selecting the right regulator for the job. Therefore, it would serve no useful purpose to deal with representative power supplies using these ICs—the material in chapter 1 more than suffices.

In addition to the direct and simple use of the three-terminal regulator by itself, various associations with other active devices often yield profitable results. For example, the tracking regulator shown in Fig. 3-5 makes use of three different active devices—a three-terminal regulator IC, an op amp, and a series-pass transistor. The negative output voltage tracks the positive output voltage, and both are approximately equally well regulated. Providing the unregulated power supply is adequate, this dual supply can deliver up to 5 A from each regulated output.

The upper half of the schematic diagram is conventional for the LM138. The lower half reveals an op-amp nulling circuit in conjunction with a PNP series-pass transistor. Note that the op amp senses the midpoint of the positive and negative output voltages. If the midpoint is positive with respect to ground (positive output higher than negative output) the op amp tends to close down the series-pass transistor, thereby increasing the negative output voltage. If the op amp senses a negative voltage at the midpoint (negative voltage higher than the positive output voltage), the op amp tends to open up the series-pass transistor, thereby decreasing the negative output voltage. In both of these corrective actions, the op amp ceases its corrective action when the midpoint of the 5.6 k ohm resistors is at zero potential with respect to ground. This condition corresponds to the desired equality between the positive and negative output voltages.

An interesting and useful aspect of this circuit is that the tracking accuracy remains good even for unbalanced loads. However, much depends upon the integ-

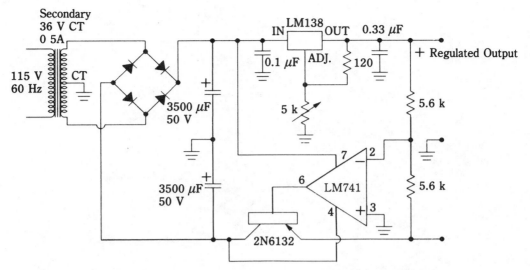

3-5 A tracking regulator using three types of active devices.

rity of the 5.6 k ohm resistors. Rather than the use composition types, it is better to use low temperature coefficient metallic-oxide resistors with 1 percent tolerance. Otherwise, this circuit is both flexible and forgiving; a wide variety of active devices other than the designated types can be made to work satisfactorily. If a grounded heat sink is used for the series-pass transistor, a mica or ceramic insulating spacer will be needed.

Synchronous rectification

Synchronous rectification has a long history, but in the past it rarely offered unique advantages for electronics applications. Recently, this technique has been revived because higher rectification efficiencies can sometimes be attained than by the use of either ordinary or exotic rectifier diodes. Synchronous rectification occurs when the control electrode of a three-terminal active device is impressed with an appropriately timed turn-on signal. The simple circuit employing a PNP germanium transistor shown in Fig. 3-6A provides quick insight into the basic concept. The selection of the old-fashioned power transistor is explained subsequently.

In the circuit of Fig. 3-6A, the arrangement is such that the transistor receives forward bias at the same time the emitter-collector circuit is properly polarized for collector current (that is, when the collector is negative with respect to the emitter). Thus, the transistor is synchronously turned on to supply current to the load for one-half of the ac cycle. This constitutes half-wave rectification. This simple circuit is not very practical, primarily because the transistor is not likely to be saturated over the full duration of the rectified wave. Next to the synchronization process, saturated operation of the transistor is very important; it brings about high rectification efficiency, and protects the transistor from destructive dissipation.

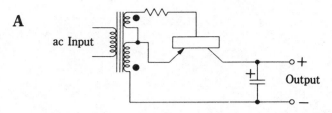

Basic circuit for demonstration.

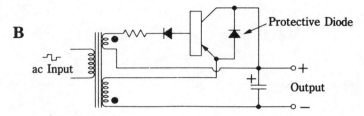

Practical circuit for high-current, low-voltage loads.

3-6 Synchronous rectifier circuits using a germanium power transistor.

Figure 3-6B shows a more practical implementation. This circuit operates with a steep-sided ac wave, which can be a square wave, a rectangular or stepped wave, or a PWM wave train. The small signal diode in the base lead prevents the emitter-base diode from being driven into the reverse-breakdown region; this diode also tends to promote quicker turn off of the transistor. The diode shown connected from emitter to collector protects the transistor if it should come out of saturation. This could be caused by inadequate ac drive, or by some load conditions. Note that a choke is not included for filtering—both inductive and excessive capacitive loading can cause penetration of the transistor SOA rating. Under normal load conditions, satisfactory operation can usually be had from just enough capacitive filtering to reduce ripple sufficiently.

The reason for specifying the PNP germanium transistor is that this almost forgotten device has an exceedingly low value of $V_{C(sat)}$. Indeed, its forward voltage drop can be much less than that of the best silicon-junction rectifier diode, and can be comparable to, or better than, that of Schottky diodes. This implementation is intended for low voltage—5 V or less—power supplies which deliver high current, say 10 to 50 A. Because germanium transistors are low-frequency devices, their application must be restricted to perhaps several kilohertz. Where applicable, however, such synchronous rectifiers might merit consideration in terms of both cost and operating efficiency. When the rectifier is initially placed in operation, the transistor remains out of saturation while the filter capacitor is charging. During this time, the protective diode is forward biased and absorbs much of the dissipation that would otherwise have to be handled by the transistor. When the capacitor is nearly fully charged, the transistor is able to go into saturation, at which time the voltage across the protective diode becomes too low for conduction. Therefore, the protective diode remains inactive unless some load or circuit disturbance again causes the transistor to fall out of saturation. This mode of operation is possible because the protective diode needs about 600 or 700 mV to conduct, whereas the voltage applied to it when the germanium transistor is saturated tends to be about 100 to 300 mV.

The major surviving manufacturer of germanium transistors is Germanium Power Devices Corp. This firm makes devices that are optimized for deployment as synchronous rectifiers, some of which have 100 A ratings.

The synchronous rectifier scheme of Fig. 3-7 provides full-wave rectification from silicon NPN power transistors. Although $V_{C(sat)}$ is not as low as that attainable from germanium power transistors, the forward voltage drops can, nevertheless, be appreciably lower than that of silicon rectifier diodes. Silicon transistors have higher frequency capabilities than germanium transistors and, of course, have thermal advantages. However, manufacturer should be queried for special power transistors intended for use in synchronous rectifier circuits. The circuit of Fig. 3-7 operates successfully with an LC filter. The secondary winding on the filter choke generates voltage pulses, which help turn off the transistors. C1 is a speedup capacitor for this function, and R1, R2 and R3 are current-limiting resistors. Diode D2 is the *free-wheeling* or *catch diode* commonly used in switching supplies and inverters. It is not as important is a full-wave configuration as it is in a half-wave configuration, but in practical systems it is still needed to transfer to the load the energy stored in the choke. D2 and D3 are small-signal diodes that help steer base signals between the two transistors. The Unitrode Corp. makes special silicon power-transistors optimized for synchronous rectifier operation.

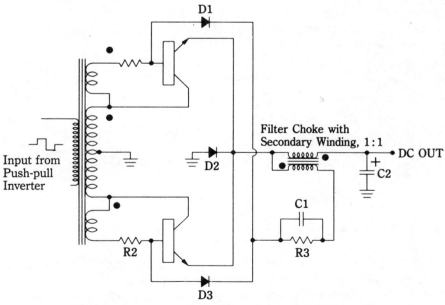

3-7 Synchronous rectifier system for push-pull inverters.

Power MOSFETs in a synchronous rectifier bridge

The use of power MOSFETs in synchronous rectifiers stems from several conditions. The fact that these devices have frequency capabilities comparable to Schottky diodes is initially an attractive feature. The fact that they can handle higher voltages is important too, because this is one of the drawbacks of the Schottky device. Until recently, the drain-source resistance of the power MOSFET has been relatively high and discouraged consideration of the device as a rectifier. However, in many applications this no longer looms as a problem and there is an ongoing trend for reduction of the resistance—what was once assumed to be an intrinsic shortcoming of this device is no longer viewed that way. Newer power MOSFETs can perform very efficiently as rectifiers because of their high-frequency response and their acceptable on resistance. Thus, they are prime candidates for switching power-supply designs, especially those involving switching rates beyond 20 kHz.

Another compelling factor for considering the use of power MOSFETs in synchronous rectifiers is that they have become available in P- as well as N-channel types. This availability facilitates circuit implementation. Also, power MOSFETs have become cost competitive with other rectifiers and rectification techniques. Of greater importance, semiconductor companies are optimizing certain power MOSFETs for synchronous rectifier service.

The bridge-configured synchronous rectifier shown in Fig. 3-8 is interesting because of its simplicity. Note that the rectifying elements comprise two N-channel and two P-channel devices. In tracing out the operating cycle of this scheme, keep in mind that power MOSFETs, unlike bipolar transistors, can conduct with reversed output-circuitry polarity. Thus, a P-channel MOSFET, which ordinarily operates with the drain negative with respect to the source, can also operate with the drain positive with respect to the source.

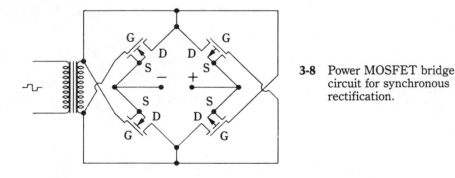

3-8 Power MOSFET bridge circuit for synchronous rectification.

It is important to use devices with low values of R_{DS} in this circuit. This is not merely to obtain high rectification efficiency. If R_{DS} is too high, the voltage drop would then be sufficient to turn on the parasitic diode that all power MOSFETs contain between drain and source. See Fig. 3-9. If the parasitic diode turns on, the

A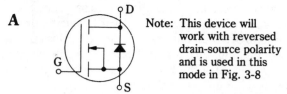

Note: This device will work with reversed drain-source polarity and is used in this mode in Fig. 3-8

Realistic symbol of an N-channel power MOSFET.

B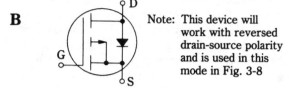

Note: This device will work with reversed drain-source polarity and is used in this mode in Fig. 3-8

Realistic symbol of a P-channel power MOSFET

3-9 Power MOSFET symbols showing parasitic diodes.

circuit of Fig. 3-8 will still perform as a full-wave bridge rectifier. However, the MOSFETs, as such, will now be out of the picture and rectification will be performed by the parasitic diodes. But such a conventional rectifier will exhibit both high conductive losses and high switching losses. An easy check to determine the mode of rectification is to measure the drain-source voltage drops. If they are less than 1 V, it is probable that synchronous rectification is taking place. Much higher voltage drops than this infer at least some undesired forward conduction of the parasitic diodes.

As in the synchronous rectifiers using bipolar transistors, the manufacturers of power MOSFETs should be consulted for information pertaining to power MOS-FETs optimized for use as synchronous rectifiers. Such devices will have inordinately low values of B_{DS}. They will thereby provide high rectification efficiency because of a low forward voltage drop. Generally, such dedicated devices also display gate turn-on voltages lower than those intended for use in conventional amplifier applications.

Power MOSFET in an electronically controlled rectifier

An ideal rectifier conducts perfectly in one direction of current flow, but totally impedes flow in the opposite direction. Practical semiconductor rectifiers require a threshold voltage to initiate current flow in the forward direction, and thereafter develop an increased voltage drop because of ohmic resistance. In the reverse direction, they might or might not limit current to a satisfactory degree—much depends on material, fabrication, and operating temperature. There are other nonideal features of these rectifying diodes; their rectifying efficiencies might show considerable degradation as the frequency is raised. Or, the revere polarity blocking ability may not hold at the voltage levels being used because of zener or avalanche breakdown.

And sometimes capacitance, nonlinearity, and noise production enter the picture. One aspect of junction-diode behavior—forward voltage drop—merits particular attention in power-control circuits because of the power loss that is thereby incurred.

Several ways of dealing with forward voltage drop have evolved. In principle, the so-called contact voltage of the PN junction can be reduced by several orders of magnitude by means of an operational half-wave rectifying circuit. Here, the 0.6 V or so across the junction is divided by the open-loop gain of the op amp, and the overall circuit performs much like a point-contact diode with a near-zero threshold voltage. This approach is fine for low-level signals, but is not practical for the voltages and currents found in power work. A second approach involves the selection of semiconductor materials. Paradoxically, the best material from the standpoint of low-threshold voltage is the once-popular germanium. To this day, the favorite advertising theme of one of the few remaining germanium semiconductor firms is the fact that large germanium rectifying diodes can perform with less voltage drop than either silicon PN diodes or silicon Schottky diodes. Unfortunately, germanium devices have limitations in other characteristics; their thermal characteristics are inferior to those of silicon devices. And germanium transistors, which show very good rectifying efficiency in low-frequency synchronous rectifying circuits, do not display the frequency capability of silicon transistors.

Schottky diodes rectify well at high frequencies and develop appreciably less forward voltage drop than do silicon diodes made with conventional junctions. Although the forward conduction behavior of Schottky diodes withstands high temperatures well, the reverse conduction at high temperatures is as troublesome as it is with germanium. Moreover, the reverse conduction degrades rapidly as a function of voltage. Although this might not always be so, Schottky diodes have a track record of being relatively costly components.

Gallium arsenide devices show much promise for simultaneous operation at high temperatures and at high frequencies, but diodes and transistors made of this material have much higher voltage drops than do silicon devices. (An example is the nominal 1.7 V drop across an LED—basically a gallium arsenide PN diode.) Nor can you go back to earlier-used materials such as selenium or copper oxide. Such materials had unique features, to be sure. Generally, they were quite leaky rectifiers, that is, they displayed a poor forward-to-reverse conduction ratio. Moreover, these earlier materials tended to lose their rectifying properties with moderate temperature rises.

From the above, note that the subject of rectification remains an important theme in power engineering. It is true that in many cases, such as rectifying from the 60 Hz power line, you can just throw in some garden-variety silicon rectifiers and your troubles are over. But in modern power technology involving frequencies into the RF region, square waves, and high-speed switching processes, improvements in rectification are very much needed. Agitating this need has been the move to 5 V heavy-current power requirements for logic and computer equipment.

A recent contribution to rectification technology involves the use of power MOSFETs. In the past, these devices have been used in synchronous-rectifier circuits and manufacturers have been optimizing MOSFETs for this dedicated purpose. A somewhat different circuit approach is shown in Fig. 3-10. This circuit is

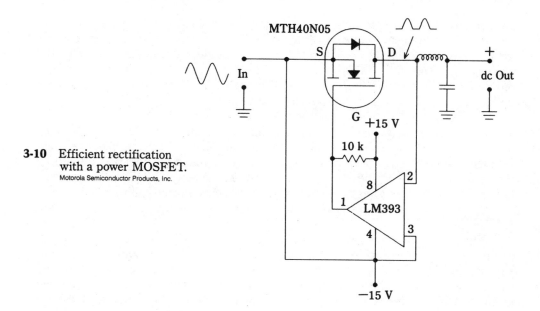

3-10 Efficient rectification
with a power MOSFET.
Motorola Semiconductor Products, Inc.

suggestive of synchronous rectification in the sense that a three-element power device is made to rectify by application of an appropriate signal to its control element. The circuit is also suggestive of precision half-wave rectifiers using op amps in the rectification process. Yet, the circuit is unique in that its operation differs from both of the aforementioned rectification techniques.

The op amp depicted in Fig. 3-10 is more specifically a voltage comparator. It functions in this circuit in such a way that whenever the input voltage is more positive than the output voltage, the gate of the power MOSFET receives forward bias and conduction takes place in this device. Otherwise, the gate is biased off and no conduction occurs in the MOSFET. Thus, the MOSFET is gated to perform as a unidirectional device, or rectifier. A natural question regarding this circuit performance pertains to the intrinsic or parasitic diode of the power MOSFET—it would appear that this diode would short out the source-drain circuit, thereby rendering the MOSFET useless in the circuit. The reason this does not occur is not obvious from the circuit; it is because the voltage drop from source to drain is much less than the ordinary contact potential (about 0.65 V) of the parasitic diode. Therefore, the parasitic diode is not given the opportunity to become active.

Because power MOSFETs have established a respectable record as power amplifiers and power switches, you might ponder why rectifying circuits such as this one did not come into prominence. The reason is that power MOSFETs were long plagued with high ON resistances. In order for the MOSFET to operate successfully in this type of circuit, the ON resistance must be low enough so that the parasitic diode will not conduct. Forward conductance of the parasitic diode is not necessarily damaging to the device, nor does it prevent the intended half-wave rectification of the circuit. However, by the time the parasitic diode conducts, the circuit has lost its principal advantages over other rectifier schemes—low forward voltage drop, and high rectification efficiency. Recently, Motorola and other manufacturers have made available power MOSFETs with both low ON resistances and greater current-handling capability than previous devices. As an example, the Motorola MTH4ON05 TMOS device used in this circuit has an ON resistance of only 0.028 ohm when delivering 10 A of rectified dc. Thus, the voltage drop is under 300 mV—better than can be achieved with a Schottky diode.

As shown, a ± 15 V low-current auxiliary source is needed for the circuit. Once the experimenter is aware of the basic principle, many different things can be done to tailor performance to individual project requirements. For example, the high-frequency performance of this circuit, as depicted, would probably be limited by the op amp long before the frequency response of the power MOSFET becomes a factor. Also, even though rectification from a sine-wave source is illustrated, operation from a square-wave source would increase efficiency and relax filtering requirements.

When evaluating this circuit, note that the voltage drop across the source-drain terminals of the power MOSFET is low, as described. If the op amp circuit is not working properly for some reason, the overall circuit could still behave as a good rectifier by virtue of the parasitic diode taking over this function. Of course, the extra margin of performance possible with intended participation of the MOSFET would not be forthcoming.

Builders of power supplies such as this one often suppose that the greatest stress on the series-pass transistor occurs at half load, full load, or in any event, somewhere in the normal operating region. Actually the power dissipation in the series-pass transistor maximizes the *foldback* region. As a rule of thumb, this tends to occur when the output voltage has fallen to about two-thirds its normal (regulated) value. That is why a liberally sized heat sink is mandatory.

Foldback action is initiated when op amp IC2 monitors a sufficient voltage drop across sensing resistance R_S to turn on transistor Q3. When this happens, the voltage developed across resistance R_C acts at terminal 1 of the CA3055 IC voltage regulator to reduce its output current. This, in turn, reduces the drive of transistor Q2 and thereby of series-pass transistor Q1. Thus, the output current of the supply is decreased as the load becomes heavier. Note that this action overrides the normal regulating action. For a full short circuit, the load current approaches, but does not quite attain, zero. In this supply, this residual load current is about 120 mA.

The fixed portion of current-sensing resistance R_S should be approximately 0.064 ohm. This could be implemented from a length of copper wire, but because of the effect of temperature, it is much better to use manganin wire. In this regard, Driver-Harris #18 manganin wire has a resistance of 0.176 ohm per foot and exhibits a very low temperature coefficient.

Note that potentiometer R20 is also involved in the current-sensing circuit. This is intended as a one-time factory-adjust provision. With R_S equal to zero (shorted out by a short length of heavy conductor) adjust R20 so that there are 200 mV between its wiper and ground. This adjustment is made under the condition of zero load current. Next, remove the shorting conductor and set the variable portion of R_S at its zero ohms position. The idea now is to determine the exact length of resistance wire in the fixed portion of R_S to allow onset of foldback operation at a load current of 3.15 A. A stable load and accurate current meter are mandatory for this adjustment. Thereafter, the variable portion of R_S will enable panel adjustment of the onset of foldback operation at load currents below 3.15 A.

The deserved popularity of the 2N3055 power transistor has been alluded to; it would only be fair to also call attention to the fact that standardization of this transistor hasn't been as rigorous as it might have been—there have been considerable differences among brands, and even among production runs of the same brand. RCA, however, prides itself on the close supervision of tight quality control of this device.

Q4 and associated circuitry form a shunt-regulated auxiliary supply for IC2 and Q3 of the foldback control section. Zener-diode CR5 protects IC1 from accidental overvoltage during test and experimentation.

Voltage-regulating power supply with current foldback

The 2N3055 power transistor has been much used as a series-pass element in linear type voltage-regulating power supplies. Its electrical ruggedness, together with useful voltage, current, and power ratings make it a natural for a wide variety of regulator designs. In order to best exploit the performance potential of this power transistor, due consideration must be given to the error amplifier, the voltage reference, and to a suitable driver stage. A very effective way of providing for these essentials is to utilize a dedicated voltage-regulator IC. These are tailor-made for the purpose; they have self-contained voltage references and have error amplifiers in the form of high-gain, but stable, differential amplifiers. A transistor driver stage suffices to interface the regulator IC with the high-current 2N3055.

Also, it makes good sense to endow such a voltage-regulated supply with a current-foldover characteristic, such that output voltage and output current *cave in* as response to a heavier than rated or heavier than adjusted for load. Such a feature protects both power supply and load from overload. It is one of the best protective techniques against catastrophic destruction of the pass transistor from a short-circuited output. Not only is it fast acting, but the supply is instantly ready for normal use once the overload condition has been remedied. Thus, the foldback mode is non-latching. The foldback characteristic is illustrated in Fig. 3-11. Here, 100-percent loading corresponds to 3.0 A, and infinite loading is a short circuit.

The schematic diagram of such a voltage-regulated power supply is shown in Fig. 3-12. (See Table 3-1 for the parts list.) From the above discussion and the block diagram of Fig. 3-13, the functional relationship of its active devices can be identified. Basically, both output of an IC voltage regulator is current boosted by a power transistor used as a series-pass element. Added to this basic configuration is a current foldback circuit that greatly reduces both output voltage and output current in the event of an overload or a short circuit. This supply is intended to provide 20 V at up to 13 A. However, it is easily modified to produce other voltage-current formats. For example, a 9 to 18 V adjustable range and a 1 A current capability

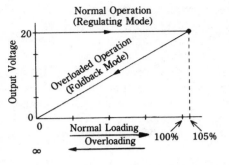

3-11 The foldback current-limiting characteristic.

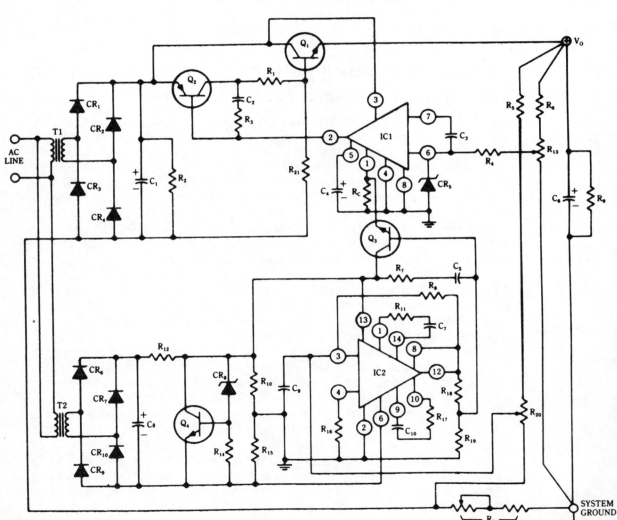

3-12 Linear voltage-regulating supply with current foldback. RCA

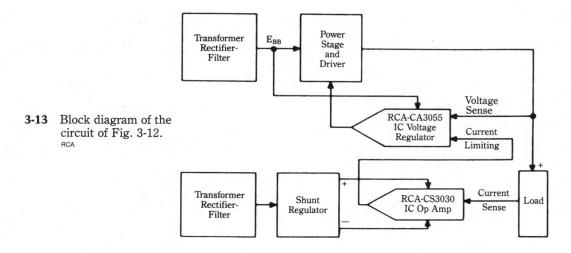

3-13 Block diagram of the circuit of Fig. 3-12.
RCA

would probably make it more useful for the general needs of solid-state power electronics. As shown in Fig. 3-12, the voltage-adjust potentiometer is intended only for a small adjustment around the nominal 20 V. (Changing the value of this potentiometer can result in an extended adjustment range for the output voltage, but it would no longer be allowable to draw 3 A unless a variac were placed in the primary circuit of transformer T1 in order to reduce power dissipation in the pass transistor.)

A word is in order with regard to the drive transistor, Q2. Although Q2 is a PNP type as opposed to the NPN pass transistor, the combination of the two form a quasi-Darlington connection. Thus, the current gain of the two transistors working together is the product of their individual current gains. Also, it is coincidental that the series-pass transistor is a 2N3055, and the voltage-regulator IC is a CA3055. The similarity in numerical designation of the two devices should not be allowed to be a point of confusion.

100 W 100 kHz switching supply

For a number of years, switching regulators have been designed to operate in the vicinity of 20 kHz. At these frequencies, audio-frequency noise from magnetic components was not a problem. At the same time, considerable reduction in size and weight was attained over lower-frequency designs. A corollary of this was that the overall power supply also became more cost effective. All this being so, designers longed for the exploitation of higher frequencies. This goal was long impeded by the general lack, or inordinate cost, of devices and components with high-frequency capability. This is no longer the case and high-efficiency regulators operating at 100 kHz and higher can be easily built.

Figure 3-14 shows the circuit of a 5 V, 20 A switching regulator that operates at 100 kHz and achieves voltage regulation by means of an IC pulse-width modulator. The configuration will be recognized as a forward converter scheme and is

Table 3-1. Parts list for linear supply with current foldback.

T1	Signal Transformer Co., Part No. 24-4 or equivalent	R12	82 ohms, 2 watts, IRC type BWH or equivalent
T2	Signal Transformer Co., Part No. 12.8-0.25 or equivalent	R13	1000 ohms, potentiometer, Clarostat Series U39 or equivalent
CR1-CR4	RCA-1N1614	R16	1200 ohms, 2 watts, wire wound, IRC type BWH or equivalent
CR5	Zener Diode, 1N5225 (3.3 V)		
CR6, CR7	Power Rectifier, RCA-1N3193	R18	510 ohms, 1/2 watt, carbon, IRC type RC 1/2 or equivalent
CR9, CR10			
CR8	Zener Diode, 1N5242 (12 V)	R19	10,000 ohms, 1/2 watt, carbon, IRC type RC 1/2 or equivalent
C1	5900 µF, 75 V, Sprague Type 36D592F0758C or equivalent	R20	300 ohms, potentiometer, Clarostat Series U39 or equivalent
C2	0.005 µF, ceramic disc, Sprague TGD50 or equivalent	R21	510 ohms, 3 watts, wire wound, Ohmite type 200-3 or equivalent
C3, C7, C10	50 pF, ceramic disc, Sprague 30GA-0.50 or equivalent	RC	240 ohms, 1%, wire wound, IRC type AS-2 or equivalent
C4	2µF, 25 V, electrolytic, Sprague 500D G025BA7 or equivalent	RS	(See text for fixed portion); 1 ohm, 25 watts, Ohmite type H or equivalent
C5	0.01 µF, ceramic disc, Sprague TG510 or equivalent		
C6	500 µF, 50 V, Cornell-Dubilier No. BR500-50 or equivalent	IC1	RCA-CA3055
		IC2	RCA-CA3030
C8	250 µF, 25 V, Cornell-Dubilier BR 250-25 or equivalent	Q1	RCA-2N3055
		Q2	RCA-2N5781
C9	0.47 µF, film type, Sprague Type 220P or equivalent	Q3, Q4	RCA-40347

R1	5 ohms, 1 watt, 1RC type BWH or equivalent

Miscellaneous

R2	1200 ohms, 1/2 watt, carbon, IRC type RC 1/2 or equivalent
R4	100 ohms, 1/2 watt, carbon, IRC Type RC 1/2 or equivalent
R5	430 ohms, 2 watts, wire wound, IRC Type BWH or equivalent
R6	9100 ohms, 2 watts, wire wound, IRC Type BWH or equivalent
R7	470 ohms, 1/2 watt, carbon, IRC type RC 1/2 or equivalent
R8	5100 ohms, 1/2 watt, carbon, IRC type RC 1/2 or equivalent
R9, R14	1000 ohms, 2 watts, wire wound, IRC type BWH or equivalent
R10, R15	250 ohms, 2 watts, 1% wire wound, IRC type AS-2 or equivalent
R11, R17	1000 ohms, 1/2 watt, carbon, IRC type RC 1/2 or equivalent

(1 Req'd)	Heat Sink, Delta Division Wakefield Engineering NC-423 or equivalent
(3 Req'd)	Heat Sink, Thermalloy #2207 PR-10 or equivalent
(1 Req'd)	8-pin socket Cinch #8-ICS or equivalent
(1 Req'd)	14-pin DIL socket, T.I., #IC014ST-7528 or equivalent
(2 Req'd)	T0-5 socket ELCO #05-3304 or equivalent
	Vector Board #838AWE-1 or equivalent
	Vector Receptacle R644 or equivalent
	Chassis—As required
	Chassis—As required
	Dow Corning DC340 filled grease

RCA

noteworthy for its simplicity, considering its respectable performance parameters. Note, in particular, the wide range of ac line voltages that can be used without any need to change input terminals. The half-power to full-power efficiency is about 75 percent throughout the ac line voltage range of 90 to 260 V.

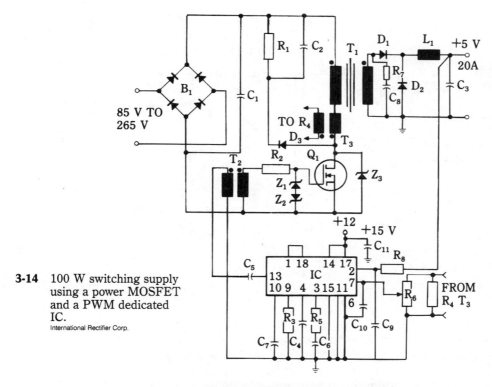

3-14 100 W switching supply using a power MOSFET and a PWM dedicated IC.

International Rectifier Corp.

REGULATED POWER SUPPLIES

Q1	1RF830 HEXFET
IC	Silicon General 3526
B1	IR 3KCB80
C1	500 μF, 450V wkg.
C2	0.68 μF, 100V
C3	4X 150 μF, 6V
C4	22 μF, 16V
C5	0.5 μ, 25V wkg.
C6	10 pF
C7	910 pF
C8	0.0068 mfd.
C9	0.005 μF
C10	0.1 μF
C11	22 μF, 25V
R1	1.5k (3500 Ω, 5 W)
R2	12 Ω 1/4 W
R3	6.8 Ω 1/4 W
R4	10 Ω
R5	12k 1/4 W
R6	100 Ω potentiometer
R7	33 Ω 1/4 W
R8	560 Ω 1/4 W

D1	20FQ030
D2	60HQ100
D3	IR 40SL16
Z1	1N4112 zener diode
Z2	1N4112 zener diode
Z3	4X 1N987B zener diodes in series
L1	Pan Magnetics International E-2481 (Core Arnold A-930157-3, 16 turns, 2 in parallel #14)
T1	Pan Magnetics International E-2478 (Core TDK 26/20. Primary: 20 turns, 3 in parallel #32; Secondary: 3 turns, 0.3 mm 0.8 cm copper strip)
T2	Pan Magnetics International E-2479 (Core TDK H5B2T10-20-5. Primary: 6 turns #24; Secondary: 6 turns #24)
T3	Pan Magnetics International E-2480 (Core TDK H5B2T5-10-2.5. Primary: 1 turn; Secondary: 100 turns #32)

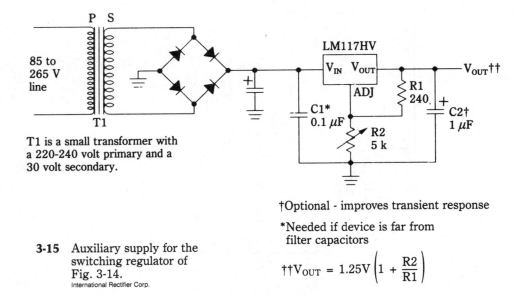

T1 is a small transformer with
a 220-240 volt primary and a
30 volt secondary.

3-15   Auxiliary supply for the
switching regulator of
Fig. 3-14.
International Rectifier Corp.

†Optional - improves transient response

*Needed if device is far from
filter capacitors

$$††V_{OUT} = 1.25V \left(1 + \frac{R2}{R1}\right)$$

Not shown in Fig. 3-14 is a 12 V power source for operation of the IC pulse-width modulator. A suitable supply for this purpose is shown in Fig. 3-15. It will provide the requisite dc operating voltage over the same line-voltage range that can be accommodated by the switching circuits. Because only 50 mA are needed by the IC under worst conditions, almost any transformer with the indicated primary and secondary windings will prove suitable for the purpose. In those applications (probably most) where the exceptionally wide excursions of line voltage do not have to be handled, an even simpler auxiliary power supply can be used; the linear regulator portion (LM117HV and associated components) can be dispensed with and the bridge rectifier can be arranged to supply a nominal 12 to 18 V of unregulated dc.

In Fig. 3-14, T1 is the main transformer of the basic forward-conversion circuit. It transfers the switched electromagnetic energy from the bridge rectifier to the output rectifier circuit. There is no clamping winding; clamping is provided by D3, R1, and C2. Transformers T2 and T3 interface the MOSFET switching circuit with the IC pulse-width modulator. Such transformer coupling is needed because of the difference in operating voltage levels between the IC and the MOSFET switching converter.

Tracking a negative voltage

In both design and experimentation it is often awkward to find that a negative voltage in the 10 to 15 V region is needed in addition to the already present positive supply within that range. Usually, a desirable alternative to building an entirely new negative supply is to reverse the polarity of the already existing positive supply. In essence, this requires a dc transformer. Such a circuit is shown in Fig. 3-16.

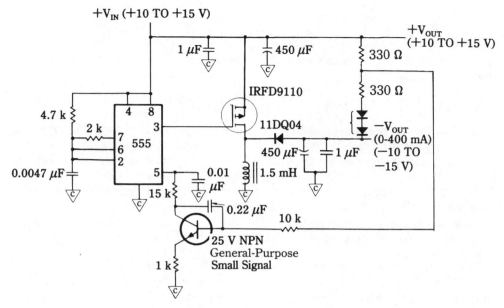

3-16 Tracking negative supply. International Rectifier Corp.

It not only will furnish the negative voltage, but will track the positive voltage supply. This is desirable inasmuch as most ICs operating from split-polarity power work best with equal-value voltages. Equal-value voltages are beneficial for other loads as well. For example, it is often necessary to preserve dc balance in voltage amplifiers and driver circuits of stereo amplifiers.

In Fig. 3-16 the 555 timer IC is connected as an astable multivibrator and would produce a 50-percent duty cycle all the time were it not for the feedback path to pin 5 through the NPN transistor. The rectangular wave output at pin 3 alternately turns the P-channel power MOSFET on and off; in so doing, the 1.5 mH (millihenry) inductor in the drain circuit is made to ring, and the negative-going alternations are passed through the diode to the 450 μF filter capacitor. Thus, although the energy derives from the original positive supply, a new dc source is now available that is negative with respect to system ground, or common.

It is the nature of the 555 timer that any imposed change in the voltage already present at pin 5 will alter the timing period. Because this IC is connected as a free-running oscillator, this manifests itself as duty-cycle modulation. In this way, the average value of the switching voltage at the power MOSFET gate can be made to follow the positive supply. More specifically, this is brought about via the voltage sensing point at the junction of the 330-ohm resistors, and via the NPN transistor in the feedback path. Note that this transistor needs no external source of collector voltage, this being provided at pin 5 of the 555 IC.

The small power MOSFET can dissipate about 1.5 W in this application; this calls for some attention to heat removal. The four-terminal, dual in-line package can convey heat from the device into the copper of a printed-circuit board. Two of these terminals are connected to the drain. Strive for low mounting above the PC (printed circuit) board, thick copper board stock, and a mounting area free from nearby heat-producing components. An improvised copper clip to fit over the device can function as a heat sink if some elementary fins are fashioned to transfer heat to the ambient air.

A unique step-down converter

Early switching power supplies and regulators used individual transistors to accomplish required circuit functions. This led to a high parts count together with reliability and performance problems. Practical implementations often did not come very close to theoretical capabilities. Part of the problem was the difficulty of attaining a circuit layout that abided with good high-frequency practice. Also, with discrete transistors, it was not an easy task to obtain consistent operation in actual production runs. A substantial improvement was realized when op amps became available for the various control functions needed in switching supplies. This led to more compact circuit layouts, to upgrading of reliability and consistency, and to more refined performance. At the same time, it helped bring costs down.

Another quantum leap in switching-supply technology came about with the introduction of pulse-width modulator ICs. With these ICs, all control functions were contained within a dedicated module, and it was only necessary to supply a power switch and a few external passive components. This made possible even higher performance levels; greater efficiencies were attained, together with tighter regulation. Also, because of the compacted circuitry, it became easier to constrain RFI and EMI. Moreover, these dedicated control modules provided useful pre-engineered sophistications, such as thermal shutdown, dead-time control for push-pull inverters, current limiting, and self-contained voltage references.

Yet another notable advance in switching-supply technology has more recently occurred. This advancement involves ICs comprising pulse-width modulator circuitry and the power-output stage on a single monolithic chip. Previously, the closet approach to this goal was achieved via hybrid technology. The Lambda LAS 6380 switching regulator series is the pioneer device making use of the monolithic format. The LAS 6380 not only contains an 8 A Darlington output transistor, but its control circuitry contains a temperature-stabilized voltage reference, internal current-limit protection, internal thermal shutdown, a dc to 200 kHz oscillator, an inhibit/enable control pin, and double pulse suppression logic.

A step-down converter circuit using the LAS 6380 is shown in Fig. 3-17 (see Table 3-2 for the parts list). This unique regulating power supply is intended to convert a nominal 20 V source to 5 V at up to 18 A. The switching rate is 40 kHz, so inductors and capacitors are of reasonable size. Because of this high switching rate, free-wheeling diode D1 is specified as a Schottky diode. For the most effective utilization of the capabilities of D1, its cathode should be as close as possible to pin 8 of the LAS 6380; its anode should be as close as possible to the ground point of the 0.1 μF capacitor in the pin 4 circuit. Ground-loop avoidance will be approached by placing C_{IN} close to pin 1. Maximized thickness should be used for printed-circuit copper runs carrying high currents.

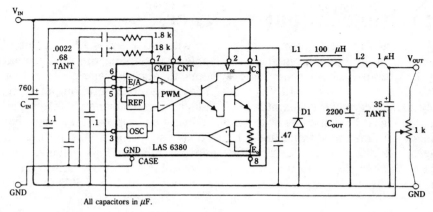

3-17 Step-down converter using combined PWM control and power switch. Lambda Semiconductors

Table 3-2. Performance characteristics of the step-down converter of Fig. 3-17.

Parameter	Conditions	Limits Min	Typ	Max	Units
Output Voltage Tolerance	$T_J = 0°C–125°C$				
V_{OUT} Trimmed to 5 Volts	$V_{IN} = 12–30$ V		±2	±3	
	$I_{OUT} = .25–8$ A				%
Output Voltage Initial Accuracy	$I_{OUT} = 1.0$ A		±3	±5	%
System Conversion	$I_{OUT} = 4.0$ A	73	77		%
Efficiency	$I_{OUT} = 8.0$ A	70	74		%
Output Noise and Ripple	$V_{IN} = 30$ V		.03		Volts Peak
			.012		Volts RMS
	V_{IN} 25 V + 5 V_{PK} @ 120 HZ		.03		Volts Peak
Output Current Limit			10.5		Amps
Output Dynamic Response Turn-On Overshoot					
Peak	$I_{OUT} = .05$ A		500		mV
Duration			40		mS
Peak	$I_{OUT} = 8$ A		0		mV
Duration			0		mS
Unit-Step Load Change	I_{OUT} from short circuit to 8 A		0		mV
Peak					
Duration			0		mS
Peak	I_{OUT} from 8 A to .05 A		500		mV
Duration			60		mS

$V_{CC} = V_{IN} = 20$ Vdc, $V_{OUT} = 5$ V, $I_{OUT} = 8$ A, Fsx = 40 kHZ $T_A = 25°C$, unless otherwise stated.

A multiple-output switcher

A particularly useful power supply is one with $+5$ V and ± 12 V outputs. This happens to be a voltage combination capable of satisfying the requirements of many TTL (transistor-transistor logic) systems, for such systems often include linear ICs that operate from ± 12 V sources. The conventional multioutput switching supply usually contains a separate choke and output transformer. The novel circuit shown in Fig. 3-18 dispenses with one of these core components. Note that the choke and output transformer are combined as a single unit. Of the three output voltages, the $+5$ V output is the most tightly regulated. This makes sense, for logic circuit operating voltage is much more critical than is dc operating voltage for linear ICs. The $+5$ V output has an output-current capability of 5 A; the two 12 V outputs can each supply 125 milliamperes—more than enough for the usual requirement for linear ICs.

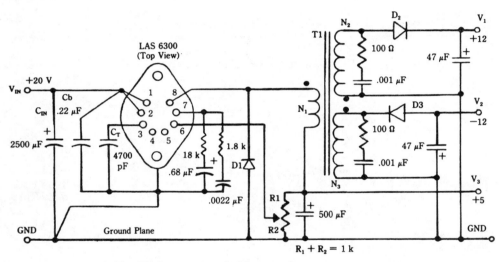

3-18   Triple-output switching supply. Lambda Semiconductors

The LAS 6300 is a sophisticated subsystem, as can be appreciated from the block diagram of Fig. 3-19. It contains its own internal linear regulator and voltage-reference source. There are also provisions for either internal or external current limiting, thermal shutdown, and remote control. Pulse-width modulation is accomplished in the comparator which samples signals from the error amplifier and the oscillator. Interestingly, current limiting is brought about by shifting the frequency of the oscillator.

This design approach takes the headaches out of switching-supply implementation because the parts count is minimal, and no external power transistor is needed. A suitable heat sink is the model 5759B-15 or the Model 5006B-15 made by AAVID Engineering, Inc. Of Laconia, New Hampshire. If due consideration is given to the phasing of the windings on T1, and Schottky diodes are used for D1, D2, and D3, no difficulties should be experienced in obtaining proper operation.

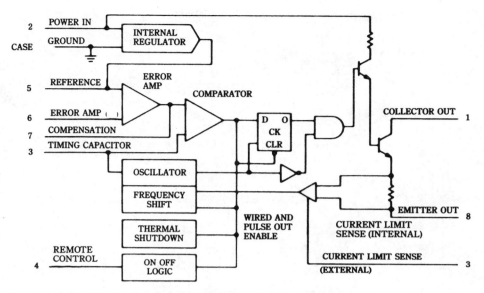

3-19 Block diagram of LAS 6300 switching regulator. Lambda Semiconductors

(This is not an easy statement to make for switching supplies comprising a large number of transistors and discrete parts.)

230 W, 50 kHz supply with multiple outputs

Switching power supplies, especially those intended for providing the requisite dc voltages for computers, generally don't qualify as simple weekend projects for the hobbyist. Yet on a comparative basis, the state-of-the-art 230 W, 50 kHz regulated supply shown in Fig. 3-20 is considerably less involved than most traditional designs. The block diagram of Fig. 3-21 will help drive home the essentially straightforward functions of the several function sections of the supply. Referring again to the schematic diagram of Fig. 3-20, note that the entire bottom half of this circuit comprises the brains, or switching logic of the supply. Further simplification can be realized by observing that the switching-logic function takes place within a dedicated pulse-width modulation IC, the Signetics NE5560. The associated blocks merely perform interface and auxiliary functions for the NE5560 pulse-width modulator.

The duty-cycle modulated buck regulator consists of the two IRF720 power MOSFETs. (These devices are called HEXFETs by International Rectifier Corp.) This regulator circuit can operate efficiently at any duty cycle between zero and almost one-hundred percent. The rectified output from the power switch then controls the 50 kHz output level of the bridge-configured power switch comprised of four IRF720 HEXFETs. This output stage operates at a constant 50-percent duty cycle, and delivers a 50 kHz square wave to the rectifier system.

Although bipolar transistor supplies have been successfully designed for

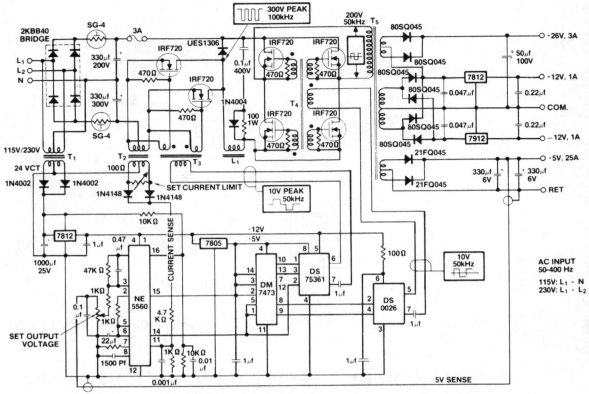

3-20 Schematic diagram of 230 W, 50 kHz regulated power supply. International Rectifier Corp.

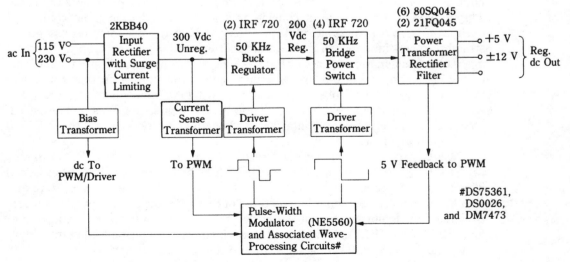

3-21 Block diagram of the circuit of Fig. 3-20. International Rectifier Corp.

50 kHz and higher switching rates, the requisite transistors border on the exotic and considerable attention must be given to snubbing circuits and other protective techniques. The HEXFETs are intrinsically capable of almost negligible switching loss at much higher rates than 50 kHz. Additionally, they are immune to thermal runaway, are easy to drive, and can be directly paralleled (no ballast resistors) for greater output power. Usually nothing needs to be done to the drive circuitry as the result of paralleling.

The salient features of the 50 kHz HEXFET switching supply are:

- There are no 60 kHz magnetics. Off-the-line operation can accommodate 115 and 230 V lines over a frequency range of 50 to 400 Hz. For 115 V operation, the power line is connected across terminal N and either terminal L1 or terminal L2. The input rectifier system then operates as a voltage doubler, producing a nominal 300 V of unregulated dc. For 230 V lines, the ac power is connected to terminals L1 and L2. The input rectifier system then operates as a conventional bridge circuit, again developing approximately 300 V dc. In either instance, the SG-4 thermistors provide soft-start characteristics for the supply.

- The output rectifier system uses no filter chokes. This is unnecessary because the 50 kHz square wave always maintains its 50-percent duty cycle. Moreover, the rise and fall times of the square wave are extremely fast. Capacitor filtering suffices for the 5 and 26 V outputs. The plus and minus 12 V outputs benefit further from the electronic filtering produced in the 7812 and 7912 linear regulators.

- Schottky rectifier diodes are used. These have low switching losses, as well as low forward voltage drops. All rectifying circuits are full wave. Note that full-wave rectification of a perfect square wave would produce pure dc with no filtering. In this supply, the HEXFETs and the Schottky diodes develop a very good square wave and filtering demands are relaxed. The 50 kHz switching rate also helps in this regard.

The feedback loop from the output is derived from the 5 V rectifier circuit. However, through the mutual coupling of the windings on T5, regulation is imparted to the other output voltage. Because the plus and minus 12 V outputs contain their individual linear regulators, this is of no consequence for these voltages. The 26 V output is, however, one that is slaved to the regulation of the 5 V output. Actually, the regulation of the 26 V output will not be quite as good as that of the 5 V output. Inasmuch as the 5 V supply to logic circuits tends to be critical in some instances, it is more important to design for maximum stabilization of the 5 V output than the others. In any event, it is generally found that 26 V sources are less demanding of tight regulation than either the 5 V or 12 V sources in the operating of computer and logic circuits.

A BIMOS bridge inverter

The inverter to be described operates from 750 V dc and delivers 10 A to a load. The switching rate is 25 kHz. As pointed out in chapter 1, such a combination of performance parameters is useful in industrial applications. However, thyristors, bipolar transistors, including Darlingtons and power MOSFETs, all exhibit profound shortcomings for service in this domain, these being cost, reliability, efficiency, or speed. By coincidence, a combination of two commonplace devices working together in a cascode circuit—the BIMOS power switch—yields performance ordinarily attainable via the use of exotic and costly devices.

The 25 kHz switching portion of the inverter comprises four such BIMOS power switches in a bridge configuration. This is shown in Fig. 3-22. It can be seen that the bipolar transistor is operated as a common-base switch, and that its emitter current can be interrupted by the MOSFET. Bipolar transistors have much better frequency and SOA ratings in the common-base circuit than in the more conventional common-emitter circuit. Because of the presence of the MOSFET, the overall power switch is easy to drive. (Conversely, a common-base power switch, by itself, would be very difficult to drive.) Note the symmetry of the bridge—all four BIMOS circuits are identical.

These BIMOS switches are a little different from the simplified circuit discussed in chapter 2. The difference pertains to the method of obtaining turn-on bias for the bipolar transistor. Initially, this bias derives from the charge stored in capacitors C1, C3, C5, or C7. However, once collector current flows, the remainder of forward base bias is obtained from the collector-current transformers, CT1, CT2, CT3, or CT4. The output of bridge converter is a nominally 25 kHz ac wave that can be controlled both in frequency and in duty cycle by the control circuit. The BIMOS bridge works well with resistive and inductive loads. (A 4 or 8 μF capacitor can be placed in series with some inductive loads to prevent magnetic saturation from dc components accompanying slight waveform imbalance.)

The control and drive circuit for the bridge is shown in Fig. 3-23. At first glance, this circuitry appears to be quite complicated. This, however is because the designer detailed the logic devices comprising the ICs. For the practical purposes of construction and testing, these ICs can be considered as *black boxes*. Such a block approach greatly simplifies the appearance of the circuit and is justified on the premise that you do not have access to the internal circuits of these ICs. On the other hand, the designer's diligence is rewarding, for it is instructional to see what is basically inside of the ICs.

IC1 is a regulator control circuit. It renders unnecessary the oscillator, voltage reference, pulse-width modulator, output stages, and other electronics commonly needed in switching-type power supplies. IC1 is the source of the two 180-degree phase-displaced waves needed for bridge gates 1 and 3. IC1 also supplies a small deadtime in these waves to protect against simultaneous conduction of the power switches. The waves needed for bridge gates 2 and 4 are then derived from these basic waveforms.

IC2 and IC4 are retriggerable, resettable monostable multivibrators. Their functions are to provide adjustable delays (phase shifts) on the edges of the waveforms for bridge gates 2 and 4. By this means, the duty cycle of the voltage waveform delivered to the load can be varied manually. Controlling the duty cycle of an ac wave in this fashion is an efficient way to control its RMS amplitude.

IC2, IC5, and Q1 through Q7 are drivers for the BIMOS power switches in the

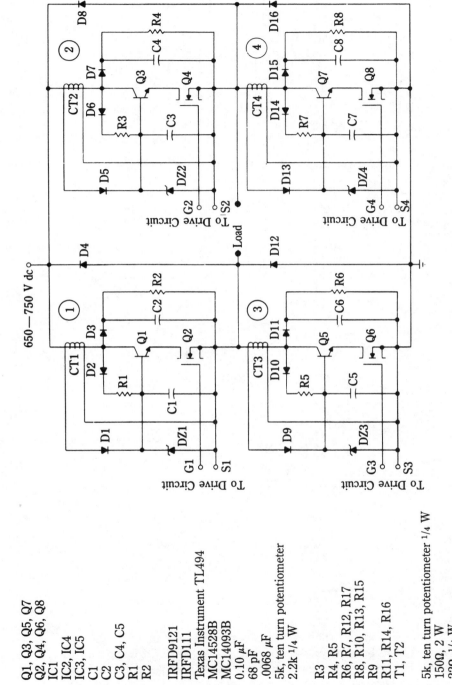

3-22 The BIMOS bridge circuit. International Rectifier Corp.

Q1, Q3, Q5, Q7	
Q2, Q4, Q6, Q8	
IC1	
IC2, IC4	
IC3, IC5	
C1	
C2	
C3, C4, C5	
R1	
R2	
IRFD9121	
IRFD111	
Texas Instrument TL494	
MC14528B	
MC14093B	
0.10 µF	
68 pF	
.0068 µF	
5k, ten turn potentiometer	
2.2k 1/4 W	
R3	
R4, R5	
R6, R7, R12, R17	
R8, R10, R13, R15	
R9	
R11, R14, R16	
T1, T2	
5k, ten turn potentiometer 1/4 W	
150Ω, 2 W	
22Ω, 1/4 W	
4.7k, 1/4 W	
500k, 1/4 W ten turn potentiometer	
50k, 1/4 W ten turn potentiometer	
Primary: 40 turns #24	
Secondary: 40 turns #24	
Core TDK H5B2T10-20-5	

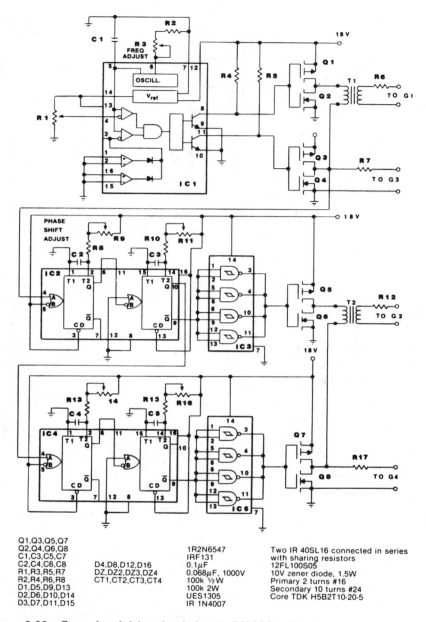

3-23 Control and drive circuit for the BIMOS bridge. International Rectifier Corp.

Q1,Q3,Q5,Q7		1R2N6547
Q2,Q4,Q6,Q8		IRF131
C1,C3,C5,C7		0.1μF
C2,C4,C6,C8		0.068μF, 1000V
R1,R3,R5,R7	D4,D8,D12,D16	100k ½W
R2,R4,R6,R8	DZ,DZ2,DZ3,DZ4	100k 2W
D1,D5,D9,D13	CT1,CT2,CT3,CT4	UES1305
D2,D6,D10,D14		IR 1N4007
D3,D7,D11,D15		

Two IR 40SL16 connected in series
with sharing resistors
12FL100S05
10V zener diode, 1.5W
Primary 2 turns #16
Secondary 10 turns #24
Core TDK H5B2T10-20-5

bridge circuit. IC3 and IC5 appear to add complexity to the operation of the circuit, but this is deceptive. These ICs are quad, two-input NAND Schmitt triggers, but they are not deployed in such a way so as to exploit their logic potential. It will be noted that the input leads are all connected in parallel, as are all of the outputs.

Accordingly, the simple function of these ICs is to act as a low-impedance source of the fast rise and fall waveform for ultimately driving two of the BIMOS power switches. Drive of the BIMOS switches takes place through pairs of MOSFETs, which must be inputted from a low-impedance source in order to preserve the steep edges of the waveform.

This inverter is easy to troubleshoot because the control and drive circuit can readily be tested separate from the bridge circuit. Assuming all ICs and active elements are poorly connected and in good working order, a possible source of malfunction could be incorrect phasing of drive transformers T1 and T2.

The nature and implementation of the current-mode power supply

The advantages of switch-mode power supplies over linear regulators are well known. Summed up, the switching technique of regulation yields supplies that are more efficient, more compact, and more economical than their linear counterparts. Also, the greater that load power, the more significant these advantages become. For a long time, the control ICs for these switch-mode power supplies operated as voltage-mode, pulse-width modulators, and were commonly known as PWM control ICs. Basically, they sampled a portion of the regulated dc output voltage and varied the duty cycle of the switching waveform in such a way that the dc output voltage remained very nearly constant. Such servo action, of course, constitutes regulation. Inasmuch as the power switch or *series pass element* is either on or off, but never in a between state, efficiency is necessarily better than in a linear supply where the series-pass element behaves as a rheostat with its attendant I^2R loss. The general idea of the voltage-mode switching supply is depicted in the block diagram of Fig. 3-24A.

A more recent development is current-mode control IC for switching-type power supplies. Actually, it performs voltage regulation in much the same way as in its predecessor, the voltage-mode control IC. In both ICs, dc-voltage regulation takes place via pulse-width modulation (that is, variation of the duty cycle of the switching wave). In both instances, a tendency for the dc output voltage to drop is counteracted by allowing the switching device to remain longer in its ON conductive state. Conversely, both types of control ICs counteract any tendency for the dc output voltage to rise by reducing the ON time of the switching device. The current-mode control IC has, however, an additional feature. It samples not only a portion of the dc output voltage, but also a voltage representing the current through the switching device. The general idea of the current-mode switching supply is depicted in Fig. 3-24B.

From the above, you can appreciate that the current-mode IC contains two feedback loops. One of these senses output voltage, as in the simpler voltage-mode control technique. The other feedback loop senses the current ramp through the switching device. The first loop can be construed to be stupid in that it attempts to regulate the dc output voltage no matter what causes it to attempt to rise or fall. This being the case, you need not be too concerned over the effect of the second

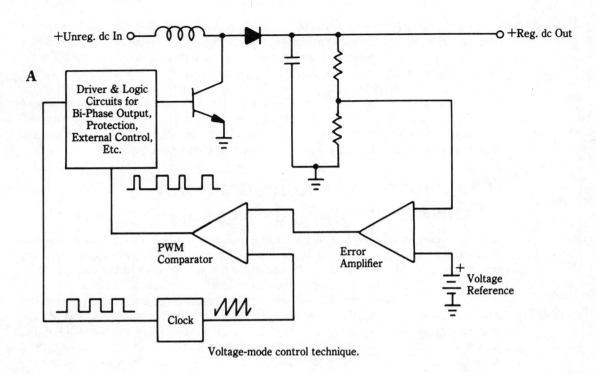

A

Voltage-mode control technique.

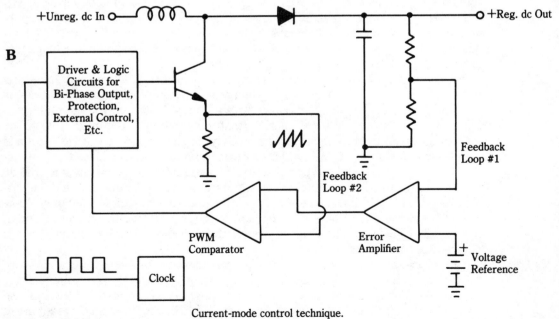

B

Current-mode control technique.

3-24 Comparison of voltage-mode and current-mode PWM power supplies. Note additional feedback loop
in the current-mode circuit.

loop on voltage regulation as a function of load changes. The current-sensing loop contributes some very worthwhile features to the overall performance of the power supply.

In the voltage-mode control technique, the PWM wave is generated by a voltage comparator working from a dc error signal and a sawtooth or triangular voltage. The sawtooth is derived from the same internal oscillator that produces the switching. In the current-mode control technique, the comparator also sees a dc error voltage and a sawtooth voltage at its inputs. In this case, however, the sawtooth is obtained from the current ramp through the switching device. Therefore, there are now two variables controlling the duty cycle of the PWM wave; one is the dc error signal, the other is the amplitude of the voltage representing the current ramp through the switching device.

Now you are in a position to appraise the performance benefit of the current-sensing feedback loop. A line-voltage transient is quickly felt as a change in the ramp current through the switching device; this action enables an appropriate response in regulation to prevent appearance of the transient at the output of the power supply. That is, the comparator is quickly alerted to the presence of the line transient. This is not the case, however, in the simpler voltage-mode control technique that lacks the current-sense provision. In this case, the line transient must travel through the LC filter of the supply before its effect can be communicated to the comparator. Unfortunately, the LC filter imparts both, attenuation and delay to the transient; by the time the comparator receives notice, the line transient has come and gone. Thus, the voltage-mode controller has limited ability to perform good line-voltage regulation. In contrast, the current-mode controller responds quickly and produces superior line-voltage regulation.

A second, not so obvious, attribute of current-mode control involves stability. In the voltage-mode control technique, the amount of usable gain in the error amplifier is generally less than desired because of the phase shift in the LC filter; high loop gain can lead to instability. Because of the varying conditions of power supply operation, compensating methods cannot be relied upon to extend gain beyond a certain point. In the current-mode controller, the effect of the filter phase shift is appreciably less. This enables higher gain to be deployed in the feedback paths so that tighter regulation can be achieved.

The practical implementation of current-mode control is easy because the dedicated control ICs can be associated with the power switch and the handful of circuit components in much the same way as the long-used voltage-mode control IC. One extra provision is needed, however. A means must be provided for sensing a voltage representing the current through the switching device or power switch. The simplest way of doing this is to sample the voltage drop across a low-value resistance placed in series with the switching device. This is straightforward enough, but it exacts a small toll on the overall efficiency of the supply. Instead of the resistance, a current transformer can be used. Such a current transformer often has a single-turn primary and, perhaps eighty or a hundred turns on its secondary winding. The current transformer avoids the I^2R loss of the series resistance, but may be objectional because of cost or space considerations. Also, some operational problems can be encountered due to leakage inductance.

A recently developed power MOSFET elegantly solves the current-sensing problem for current-mode supplies. A conventional power MOSFET, such as would ordinarily be used as the switching device, actually comprises thousands of tiny FET *cells* in parallel. If one or several of these devices is brought out to a fourth terminal, the current available will be a small fraction of the current delivered by the majority of the cells in the device. You can then connect a quarter-watt resistance between the several cells and ground and the voltage drop across this resistance will mirror the very much larger current passed by the thousands of paralleled cells. The separate lead connecting to the small group of cells is called the mirror or *pilot* terminal. As these terms suggest, this arrangement enables the sensing of the large current in the drain-source circuit. Appropriately, Motorola calls these power MOSFETs *SENSEFETs*.

The current-mode power supply shown in Fig. 3-25 is designed around a SENSE-FET. This implementation dispenses with the current transformer or the source resistor needed for sensing current with conventional power MOSFETs or bipolar transistors. Unlike a resistor in the source circuit, the resistor shown in the mirror circuit consumes negligible power, and thus does not degrade operating efficiency. Otherwise, the basic operating principles of the current-mode supply are preserved.

Although the SENSEFET power device features four basic elements (gate, drain, source, and mirror, or pilot), most of these devices provide for five pin or lead connections. This is not obvious in the circuit of Fig. 3-25; it is often all right to use just the four pins corresponding to the four elements mentioned.

The fifth pin is known as the *Kelvin connection*, and is designated by the letter *K*. The Kelvin connection is essentially another lead brought out from the source. By its use, the heavy drain-source current does not contaminate the relatively feeble current in the mirror circuit. (Typically, the sampled current in the mirror circuit is about 1/2000 of the load current in the drain-source circuit.) Use of the Kelvin connection improves the accuracy of the mirror current as proportional representative of the drain-source current. An ordinary ohmmeter would show substantially zero ohms between the source and Kelvin pins. Nonetheless, the current-sampling accuracy of the device suffers somewhat if the gate-source and the drain-source circuits are linked by the common impedance of the bonding wire between the source pin and the actual source region within the device itself.

The Kelvin connection is more clearly shown in the symbol of the SENSE-FET, as depicted in Fig. 3-26. Note the five-pin connections in the device package shown in B of Fig. 3-26. In the event that neither the mirror nor the Kelvin pins are used, the SENSEFET would operate in a power-control circuit exactly as a conventional power MOSFET.

Light-load performance boost for PWM regulators

It often happens that a PWM regulated power supply exhibits poor performance at light loads. This might assume the form of instability, excessive ripple, poor regu-

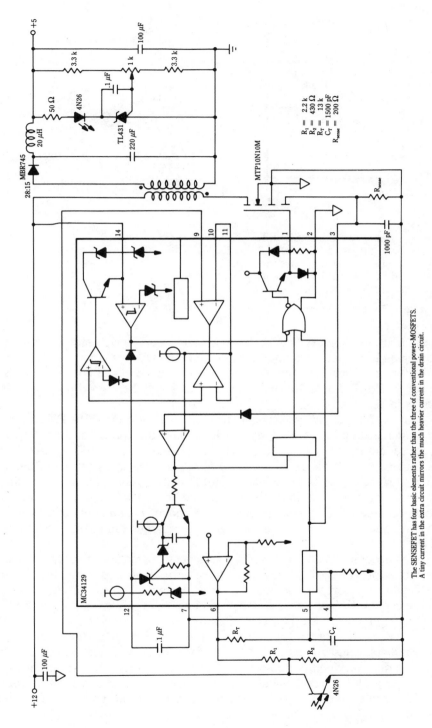

The SENSEFET has four basic elements rather than the three of conventional power-MOSFETS. A tiny current in the extra circuit mirrors the much heavier current in the drain circuit.

3-25 A current-mode power supply using a Motorola SENSEFET. Motorola Semiconductor Products, Inc.

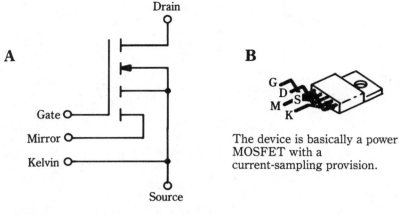

A

Drain

Gate

Mirror

Kelvin

Source

Symbol of the SENSEFET.
Note the extra (Kelvin)
connection brought out in
addition to the conventional
source-connection.

B

The device is basically a power
MOSFET with a
current-sampling provision.

3-26 Symbol and pinning arrangement of the SENSEFET. _{Motorola Semiconductor Products, Inc.}

lation, or audible sounds despite the 20 kHz or greater switching rate. Not much is to be found in the technical literature dealing with this situation because it is generally assumed that operation will be at half to full load rather than 10 or 20 percent of full load. There are, however, applications where good operation at light loads is important.

Half-wave rectification is usually used in order to convert the PWM waveform to dc. Such a circuit is particularly sensitive to the inductance of the filter choke in the sense that ever larger inductance is needed to maintain steady current flow as load current is decreased. On the other hand, a relatively low value of inductance suffices for the region between half- and full-load current. This leads to a dilemma if the supply is to operate under both light and heavy load conditions. If an attempt is made to design a suitable high-inductance choke, you quickly run into contradictions imposed by core saturation, wire size, weight and dimensions, to say nothing of cost.

A compromise suitable for many practical operational situations is usually attained by providing the core of the choke with an air gap. Although the air gap sacrifices use of some of the core magnetic permeability, saturation is prevented, and the choke exhibits a near-constant inductance over a wider load current range than would otherwise be the case. Often, however, this technique still fails to satisfy the needs of light-load operation. Some means must be found to enable the choke to display high inductance to low load currents; such a choke could, however, be less inductive when exposed to higher load currents. Such requirements are suggestive of the swinging choke once popular with class B electron-tube amplifier rectifier circuits. For the PWM regulator, a more abrupt transition from high to moderate inductance can be accommodated.

Having identified the nature of the filter-choke problem in PWM regulators, you can see why such a power supply is particularly vulnerable to the cessation of steady current which comes about at light loads. Recall that such a supply attempts to restore a rising output voltage by narrowing the ON duration of the PWM wave. Also, the primary reason output voltage tends to change light loads is that the choke inductance is no longer sufficient to sustain uninterrupted current through it. A corollary of this is that the filter then behaves as an input inductance capacitor circuit, rather than as an input inductance circuit. This being the case, the output voltage tends to change. Admittedly, the situation is not as drastic as in sine-wave circuits where the voltage rise would be about 40 percent. However, transients and ripple become more pronounced and make more severe demands on the error amplifier. Ideally, the regulation process should counteract these and other effects tending to change the dc voltage delivered to the load. Practically, there is a limit to which the ON pulse of the PWM wave can be narrowed.

When the ON pulse of the PWM wave is wide, the effects of finite rise and fall times are negligible; such pulses appear perfectly rectangular. But, if these pulses are allowed to become too narrow, the rise and fall times become appreciable portions of the pulse duration, and there is a significant change in pulse shape. This change can produce various disturbances in the regulation process. Even in situations where rise and fall times are very fast, too narrow ON pulses force the error amplifier into regions straining its response capability. All things considered, it is desirable that uninterrupted current flows through the filter choke regardless of load. This objective can be closely approached by the use of a nonlinear choke that provides high inductance at light loads.

Figure 3-27 shows the effect of small and large air gaps in the cores of conventionally designed filter chokes. Neither situation can accommodate a wide range of load currents. Small gaps lead to early saturation as load is increased. Large gaps do not allow enough inductance for smooth operation at light loads. And a medium gap would still force compromise at load extremes. So much for ordinary magnetic gaps.

It turns out that a more advantageous variation in inductance can result if a sectionalized, stepped, or tapered gap is used. The idea is to cause delayed or non-simultaneous saturation of different parts of the magnetic circuit. Ferrite E, or cup cores are good candidates for the experimentation needed to obtain the desired inductance versus current characteristic. Figure 3-27B shows the choke behavior from an appropriately stepped gap, and Fig. 3-28 illustrates the configuration of such a stepped gap.

The electronic correction of power factor

An often overlooked aspect of electrical equipment operation is the power factor. Equipment operating at 90 percent efficiency sounds like a worthwhile accomplishment, and so it might be. However, if the specifications boast of 90-percent efficiency but say nothing of the low power factor, you do not have efficient performance. From the viewpoint of the utility company, 90-percent stated efficiency coupled with a power factor of 0.5, is tantamount to an actual efficiency of about 45 percent. This is because the current consumption of equipment operating

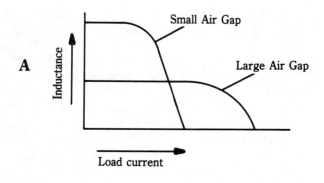

A

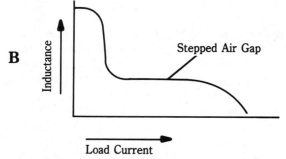

B

3-27   Effect of air gaps on choke inductance.

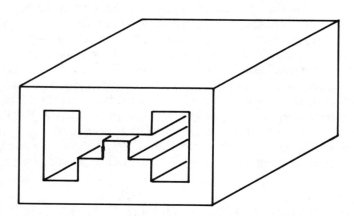

Desirable inductance behavior can be obtained by this technique. Gap geometry and dimensions are best determined empirically.

3-28   A stepped gap in the core of a filter choke.

at 0.5 power factor is about twice what it would be for unity power factor. Thus, despite the 90-percent claimed efficiency, fuses, circuit breakers, conductors, and transformers are subjected to unnecessarily high stress.

Unfortunately, much equipment tends to subject the ac line to a low power factor. Phase-control circuits using SCRs and triacs are especially offensive despite the fact that no appreciable inductance or capacitance is involved. Texts on electrical engineering often emphasize the effects of inductance and capacitance on power factor and deal with remedial techniques. These books tend to downplay or even omit the effect of nonsinusoidal current waves, such as are produced by SCRs and triacs. This is understandable because classical electrical theory was formulated long before it became commonplace to control power by chopping current. The fact remains that such chopping causes reactive effects even though physical inductance or capacitance is negligible or absent. Such reactive effects is manifested by departure from unity power factor even though the phase-control circuit feeds a purely resistive load.

An even more common source of low power factor is the rectifier-filter input circuits of power supplies, converters, and inverters. The commonly encountered bridge rectifier and single-capacitor filter produces a distorted ac current wave, and the resultant power factor is usually in the vicinity of 0.65. The textbook correction technique inserts an inductance in the ac line. However, this approach is neither practical nor economical. What is needed is an electronic method of restoring sinusoidal shape to the ac current wave and making it occur without phase displacement from the impressed line voltage. Such manipulation of the ac current would restore operation to unity power factor, and the situation would be as if the ac power were being delivered to pure resistance.

A side benefit of such an electronic power factor corrector would be the reduction of EMI and RFI contained in the distorted current wave of ordinary rectifier-filter circuits. By eliminating line-current harmonics, such power-factor correction could be expected to reduce interference to other equipment on the line, or in the vicinity. Relevant wave shapes are shown in Fig. 3-29.

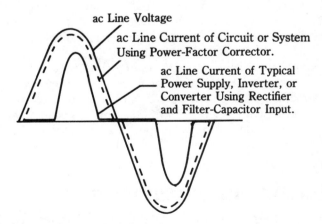

The peaked ac line current, which typifies power supplies, inverters and converters represents a low power factor. The power-factor corrector circuit (also called an active *harmonic filter*) restores the line current to a sine-wave in phase with the impressed line voltage.

3-29 Effect of using power-factor corrector.

Control ICs for correcting power factor are available. Such ICs are essentially current-mode PWM controllers with some modifications. Thus, the power-factor controller can serve as part of a switch-mode regulated supply. As such, the supply would operate from a nominal 120 or 240 V ac line and would deliver 380 V dc. It might seem that this high output voltage would be quite restrictive. However, lower dc voltages can be readily obtained by following the power-factor corrector with another switch-mode supply—one designed to work from 380 V dc and deliver the desired lower voltage.

At first thought, the notion of two power supplies in tandem appears awkward and costly. Also, you would initially suppose that the overall efficiency of such a duo would necessarily be less than could be readily attained from a single supply. Actually, the situation is quite otherwise. The power-factor corrector operates at about 95-percent efficiency and, for practical purposes, its power factor is close to 1.0. It turns out that inserting the power-factor corrector supply ahead of the ordinary supply appreciably reduces the ac line current. This is because the ac line no longer has to supply the 0.6 power factor power input of the ordinary supply. (If the ordinary supply were used directly off the line, it would have to resort to a rectifier-filter circuit to convert the ac to dc. It is this circuit that degrades the power factor.) Comparison of the two situations is shown in Fig. 3-30A and 3-30B.

From the foregoing, it is true that the overall efficiency of the two supplies must be less than that of either. But, because of the very high efficiency of the power-factor corrector, and principally because of the great reduction in ac line current from the improved power factor, the overall efficiency of such a system is increased in a worthwhile way. In practice this means that more dc output power will be available from a given installation of fuses, circuit breakers, conductors, and associated electrical hardware. Some initial expense will be involved, but this is essentially limited to the IC and its associated power MOSFET. A nice thing about such an implementation is that the dedicated IC eliminates countless design attempts, cut and try, and the inevitable catastrophes that accompany such approaches. The block diagram of the power-factor correction stage is shown in Fig. 3-31.

The ML4812 power-factor controller is made by the Micro Linear Corporation. Other vendors make similar ICs, in which instances they are sometimes called *active harmonic filters*. Note that a filter capacitor is not used following the bridge rectifier in the practical circuit of Fig. 3-32. This is because the IC must sample the sine-wave alternations of the incoming ac. Filtering is accomplished at the dc output of the supply.

Three-terminal voltage regulator

The three-terminal voltage regulator has deservedly enjoyed widespread popularity in all types of electronics applications. These devices have greatly simplified the implementation of voltage regulation, and have at the same time dramatically reduced parts count and overall cost. And by making practical the concept of regulation close to the circuitry where needed, improved performance has been real-

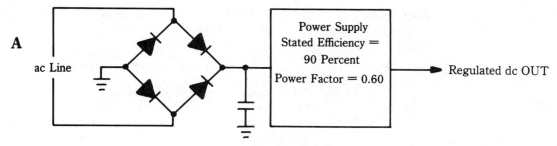

A

ac Line

Power Supply
Stated Efficiency =
90 Percent
Power Factor = 0.60

Regulated dc OUT

Practically Realized Efficiency = 0.9 × 0.6 = 54 Percent

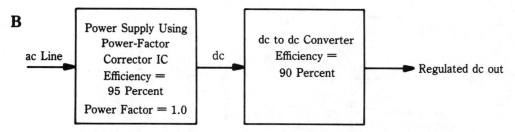

B

ac Line

Power Supply Using
Power-Factor
Corrector IC
Efficiency =
95 Percent
Power Factor = 1.0

dc

dc to dc Converter
Efficiency =
90 Percent

Regulated dc out

Practically realized efficiency = 0.95 × 1.0 × 0.9 = 85.5 Percent

Despite the use of two tandem connected devices in situation B, the operating efficiency is higher than with the single off-line supply depicted in A. This difference stems from the improved power factor seen by the ac line in B.

3-30   Improved overall efficiency via use of power-factor corrector.

ized in many different equipments. The ac-bypassing action of these regulators has greatly alleviated problems with interaction between stages and circuit sections because of the mutual impedance of unregulated dc sources, and of connecting cables. For the most part, these IC regulators have carried current ratings of between one and eight amperes, and have been available in output voltage ratings of between 5 and 35 V. Both positive and negative types have been marketed. Some, instead of having fixed output voltages, are designed to be adjustable.

The three-terminal voltage regulator was the first widely produced power IC. As such, it paved the way for other useful ICs combining brains and brawn on a single monolithically integrated chip. Although the three-terminal regulator was initially intended to be used for its specifically designated circuit function, its inherent versatility became evident early. Thus, it became common for manufacturers to show how these devices could be used in other than a narrowly specified way. For example, they are readily made to operate as current regulators; as such, they can provide either fixed or variable currents to a load. By associating them with an external power transistor, their current, voltage, or power capability can be

Feedback Loops: I. Voltage-sensing loop. This circuit senses and corrects the dc output voltage, thereby producing basic voltage regulation. It is much the same in all switch-mode systems.
   II. Current-sensing loop. This circuit is used in current-mode supplies. Its primary function is to enhance regulation of the supply against variations of line voltage.
   III. Sine-wave sensing loop. This additional loop used in power-factor correctors (active harmonic filters) forces the ac line current to follow the form of the sine-wave line voltage.

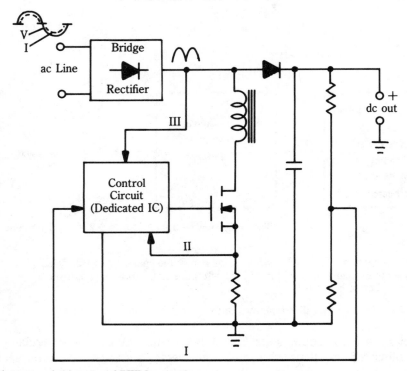

The features of this type of PWM control are:

a. No filter capacitor is used with the bridge rectifier.
b. System resembles current-mode regulators, except that line-frequency ac is introduced as modulating signal of PWM wave train.
c. Power stage is a boost regulator because dc output voltage must be higher than peak value of ac line voltage.

3-31   Basic power-factor corrector arrangement.

extended. By appropriate circuit techniques, a positive regulator can be used to regulate a negative output voltage, and vice versa. Even though best results are achieved in variable-voltage regulation by using a regulator designated for such service, a fixed-voltage type can be made to perform acceptably for many applications. Finally, these linear regulators work exceedingly well as switching regulators in relatively simple circuits.

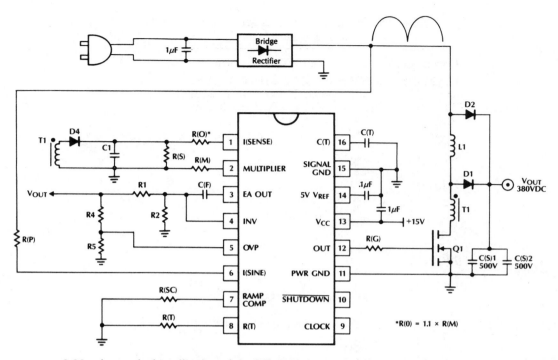

3-32 A practical application of the ML4812 power-factor corrector. Micro Linear Corporation

Despite their apparent simplicity, many three-terminal regulators contain up to several tens of active devices and provide operational features previously found only on deluxe laboratory supplies. These include thermal-overload shutdown, SOA protection of the series-pass element, current limiting, and stable temperature-compensated voltage references. Regulation specifications are often tighter than needed; line regulation is frequently found to be in the .005–.01 percent per volt range and load regulation is often specified in the .05–0.1 percent per volt range. Ripple rejection, which indicates the ability of the device to act as an electronic filter, is usually in the 60–80 db range—higher than is easily attainable in series-pass regulators designed around discrete devices.

Because the three-terminal voltage regulator is a self-contained subsystem, it might be supposed that it is only necessary to connect it between the unregulated dc source and the load and assume it need be given further thought. This is almost true most of the time; the maker has done his utmost to make these devices idiot-proof and quickly applicable to a wide variety of situations. However, success in the practical implementation of the three-terminal regulator is best achieved by awareness of its vulnerabilities and by the observance of some basic ground rules.

The simple three-terminal regulator circuit is shown in Fig. 3-33. An even simpler implementation sometimes results from eliminating capacitors C1 and C2. In general, however, these capacitors are either essential or desirable. Input capacitor C2 is often prescribed as a ceramic or solid tantalum type in the capacity range of

**Table 3-3. Component values for circuit
of Fig. 3-32.**

Component	Value/Description
L1 (2 mH)	SPANG 58076-A2 180T #24 AWG
or	Micrometals T184-40 120T #24 AWG
T1	SPANG F41206-TC 80T #30 AWG
R(S)	100 Ω
R(M)	28.8 kΩ
R(O)	33 kΩ
R1	360 kΩ
R2	4.8 kΩ
R(P)	750 kΩ
R4	360 kΩ
R5	4.6 kΩ
R(SC)	33 kΩ
C(T)	1000 pF
R(T)	14 kΩ
R(G)	10 Ω
C(S)1	1.5 μF
C(S)2	340 μF, 500V
D1	FRP850 or MUR850
D2	1N5604
Q1	IRF840 or equivalent
C(F)	0.44 μF
D4	IN4148
C1	1000 pF, Ceramic

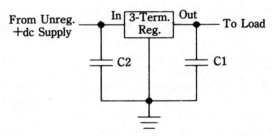

3-33   Implementation of the three-terminal regulator.

0.1 μF to several microfarads. This capacitor must be physically situated as close as possible to the regulator. In those instances where it is permissible to dispense with C2, it is because the connecting lead from the filter capacitor of the unregulated dc source is short. The prime function of C2 is to make the regulator stable, and particularly to prevent high-frequency oscillation. For this reason, aluminum electrolytic capacitors such as are commonly used for power-supply filtering are not always trustworthy. As a rule of thumb, the aluminum electrolytic capacitor would have to be 25 times or more the size of a ceramic or solid tantalum capacitor for equivalent high-frequency bypassing effect. Even the low-frequency impedance of the aluminum filter capacitor can be inadequate for satisfactory bypassing; this is because of the ESR (effective series resistance) of these capacitors.

Although ceramic capacitors are universally used for bypassing higher frequencies, some types seem to suffer an appreciable decrease in capacitance around a half megahertz, or so. Although this behavior will rarely cause trouble, it occasionally is the source of mysterious instability. All things considered, the solid tantalum capacitor is a good choice for C2. Where size considerations do not pose problems, mylar capacitors are also excellent.

Output capacitor C1, also helps stabilize the regulator against oscillation or instability. Generally, however, its prime function is to improve the transient response. It, too, can be dispensed with under some conditions. If, for example, the load has such a capacitor and is located close to the regulator, C1 might not be needed. If the regulator exhibits stability and transient response is not an important factor, C1 would not be needed. Some regulators tend towards instability when C1, or its equivalent, is relatively small—in the 500 pF to 5000 pF range. Thus, C1 is commonly stipulated to be in the 0.25 μF to several microfarads range. Here, again, ceramic and solid tantalum capacitors, as well as mylar types generally serve well.

Three-terminal, adjustable-voltage regulator

The basic circuit for obtaining adjustable output voltage from a three-terminal voltage regulator is shown in Fig. 3-34. This circuit will work satisfactorily in some applications with many of the fixed-voltage types, but best performance will be achieved with specially designated adjustable regulators, such as the LM117, LM137, LM138, and LM150 family of regulators. Typical values for R1 lie in the 100 to 300 ohm range; R2 is commonly a 5 k ohm potentiometer connected in rheostat fashion. For increased ripple rejection, the optional capacitor, C3, may be added; 10 μF aluminum electrolytic types generally serve this purpose, but heed should be paid to polarity. If this capacitor is used, the regulator becomes vulnerable to damage from certain fault conditions—this will be discussed subsequently.

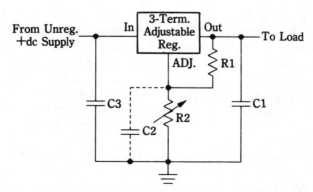

3-34 Implementation of the three-terminal adjustable regulator.

This circuit will not allow adjustment all the way down to zero voltage. However, a minimum output of 1.2 V is usually attainable and is low enough for most practical use of an adjustable voltage source.

In order to be able to adjust to zero voltage, an arrangement such as depicted in Fig. 3-35 must be used. Here, the ADJ terminal of the regulator connects to a negative 1.2 V source instead of ground. The LM113 can be considered, for practical purposes, a synthesized 1.2 V zener diode. An additional requirement is an auxiliary low-current negative 10 V source. As shown, this modification allows a 0–30 V adjustment range of the regulated output voltage.

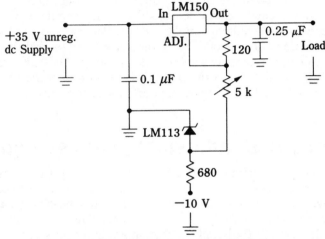

3-35   Technique for obtaining adjustable voltage down to zero.

Using diodes to protect
three-terminal regulators

Regulator ICs are vulnerable to damage or destruction from certain fault conditions. Consider the circuit of Fig. 3-34. If the input is shorted to ground, capacitor C1 will discharge into the output terminal of the regulator. Whether this portends danger to the regulator depends on the way the internal circuitry of the regulator is fabricated, on the size of C1, and on the voltage across C1. Generally, a high-current regulator that delivers low output voltage, together with a small output capacitor represents a combination of conditions in which the regulator is not endangered by such a fault. A combination of opposite conditions can, however, endanger the regulator.

Somewhat similarly, if either the input or output is shorted to ground, ADJ bypass capacitor C2 discharges into the ADJ terminal and can cause catastrophic destruction. Here again, much depends upon the size of the capacitor, the nature of the IC, and the voltage across the capacitor. The 10 μF capacitor often used in this circuit is not large enough to endanger many IC regulators, especially if 5 V or so is the output voltage. Sometimes, for the sake of increased ripple rejection, capacitor C2 is made much larger than 10 μF, whereupon its sudden discharge current can pose a threat to the regulator.

It is possible to protect against the possible consequences of such circuit faults by adding diodes, as shown in Fig. 3-36. Here, diode D1 protects against discharge from capacitor C1; diode D2 protects against discharge from capacitor C2. These diodes can be ordinary high-current types such as the 1N4002. Note that the diodes are so polarized as to exert no effect on the normal operation of the regulator.

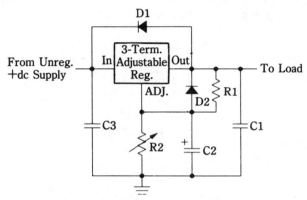

3-36   Three-terminal regulator circuit with protective diodes.

Easy way to zap the IC voltage regulator

There remains yet another way in which three-terminal and other IC regulators can be destroyed, as the saying goes without half-trying; that is, by interrupting the ground or ADJ connection. When this is done, the output voltage rises to approximately the input voltage from the unregulated dc supply. This, in many instances, damages or destroys devices or components in the load circuit. And, when the open connection is again completed, the IC regulator itself can suffer destruction. An obvious reaction to this information is, of course, that there is no reason to expect such a fault condition—that such an occurrence has a very remote probability of taking place.

Superficially, this reasoning appears sound. However, it neglects consideration of a very common test technique during breadboard evaluation, and also during troubleshooting procedures. This entails either the removal and plugging in of the regulator IC, or of the card on which it is mounted, while power is on. It doesn't require much imagination to visualize that all three terminals of the regulator, or the corresponding pins of the card connector probably do not make simultaneous contact during the plugging-in act. Accordingly, you can expect that sometimes the ground or ADJ connection will be the last in the contacting sequence. This being the case, the precaution against mysterious trouble of this nature is simple enough: **Do not remove or replace either the regulator IC or the card on which it is mounted while power is on!**

Wiring hints for three-terminal IC regulators

Because the current being processed is dc, it is commonly held that it makes little difference how the three-terminal regulator is wired or connected from a geometrical standpoint. It is, according to this notion, much like a doorbell circuit—about the same operation will ensue no matter what goes where as long as the basic current path is correct. This is far from the truth with regard to three-terminal regulators. Because of the high currents, high amplification, and close regulation

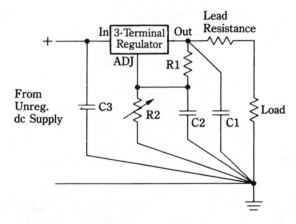

Correct method. Single-point connection for ground paths avoids poor performance from mutual ground loops.

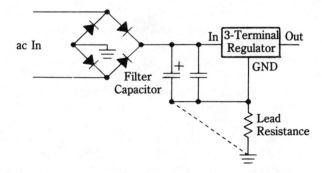

Wrong method. Filter capacitor in unregulated supply should connect to the ground point (dashed line), not to the GND terminal of the regulator.

3-37   Wiring methods can affect performance of three-terminal regulators.

expected, the tiny resistance of short lengths of connecting lead can make itself felt in performance. Indeed, more often than not, the true capability of these ICs is not realized because no effort is made to optimize the wiring pattern. Even though a designer and a technician can accept the connection diagram of a regulator, the measurements later made of actual performance might mysteriously fall well below both specifications and expectations.

In the semipictorial diagram shown in Fig. 3-37A, two very important construction techniques are depicted. First, a single ground point is used. This ground point should be as close as possible to the load. This technique avoids ground loops which easily couple from one section of the regulator circuit to another, and often cause excessive output ripple, instability, and degradation of regulation. Second, R1, the set resistor, should be connected as close to the output terminal as possible. If this connection, instead, is made near the load, the effect of lead resistance will be multiplied and load regulation can be seriously degraded.

In Fig. 3-37B the lead resistance of the connection between GND and ground is often ignored and the filter capacitor of the unregulated dc supply is connected directly to the GND terminal of the regulator. Doorbell-circuit psychology would suggest that this is the same as making this connection to ground, as indicated by the dashed line. This is not so, however; if the connection is made via the wrong method, it is possible to induce considerable ripple into the load circuit. An attempt to remedy this type of poor performance by increasing the size of the filter capacitor tends to actually aggravate the ripple in the output of the regulator.

If, because of practical reasons, the ideal connection patterns cannot be attained, they should be approached as closely as is feasible. It is always helpful, too, to make the critical lead resistances very low; this can be done by means of large conductors or multiple conductors. For example, some regulators use the TO-3 package, with the case being the GND connection. With such regulators, it is convenient to provide two paths from case to circuit ground by attaching lugs to both mounting screws.

4

Control of electric motors

INHERENTLY INTRIGUING IS THE CONTROL OF ELECTRIC MOTORS VIA THE brains and nerves of electronic devices and circuits. In such methodology, the muscle, represented by the motor, obediently follows the operating instructions conveyed by servo, feedback, or digital-programming techniques. Many basic ideas underlying such control date back to an earlier era. However, electron tubes just didn't have the right combination of electrical and physical characteristics to qualify for widespread implementation in motor-control systems. With the advent of the first power transistors, interest was revived in this field, for then there was an electronic rheostat with the kind of parameters suitable for motor control. Much was accomplished despite the limitations and inadequacies of the early transistors. Obviously, the wheels of progress had to turn for a number of years until the capability and reliability of power transistors finally merited recognition as really practical motor-control elements. While such progress was in effect, the SCR made its debut and quickly caught on as a valuable adjunct in motor-control systems. Today, these semiconductor devices and their derivatives provide an extensive family of motor-control elements with previously unthinkable ratings in voltage, current, power, and switching speed.

The motor-control applications discussed in this chapter deal with both dc and ac motors. The circuits tend to be flexible in that the scaling techniques needed for larger, smaller, or somewhat different motors than those designated are quite straightforward. This is very much because of the wide availability of semiconductor power devices with an extensive range of parameter ratings. It also follows from the accommodating nature of many control techniques. For example, power MOSFETs often can be paralleled with no need to provide increased drive power. And often a larger SCR or triac can be substituted for the stipulated one with retention of the original trigger circuit.

Motor control is an exceedingly important field; it bears relevantly on innumerable industrial processes, on traction vehicles, on automotive and aircraft auxiliary functions, and on the technology of robotics. There are also many applications involving toys and various hobbies.

Last, but not least, is the interesting phenomenon of electric control of motors with regard to motor behavior. The classic characteristics pertaining to the various types of electric motors are often drastically modified; thus, the textbook relationships linking speed, torque, power, and starting performance can be deliberately or inadvertently altered in control circuits using various sensing and feedback techniques. In general, this expands the flexibility and reduces the limitation of motors.

A variable-speed dc motor controller

The dc motor has long been known for the ease with which speed control can be implemented. All that needs be done is to vary the voltage applied to the armature (assuming a permanent-magnet type). Traditionally, this has been accomplished by means of a rheostat. The shortcoming of this method is that the overall system is very inefficient because of the power dissipation in the rheostat. Note that the same drawback prevails when the variable-resistance rheostat is replaced by a solid-state rheostat, such as a transistor. A more efficient approach is to chop the dc and obtain the variation in applied voltage by varying the duration of the ON time of the chopper. The principle of operation for doing this with power transistors is old hat, for it is merely an adaptation of the pulse-width modulation concept long used in switching power supplies. However, this approach has often suffered from inordinate circuit complications.

A relatively simple PWM speed control for DC motors is shown in Fig. 4-1. This circuit is simpler than might be assumed from first glance, for the four op

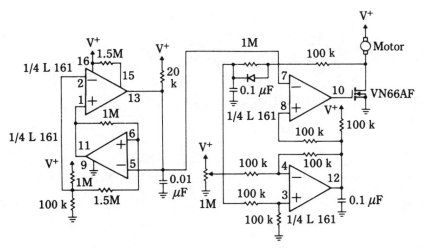

4-1 Efficient circuit for controlling the speed of a dc motor. Siliconix Inc.

amps are all part of a single IC module. All that is needed is the IC, the power MOSFET, and a few passive components. Previous implementations of this control concept usually required a tachometer. None is needed in this circuit, however.

Speed sampling is accomplished by monitoring the counter EMF (electromotive force) of the motor. This sensing is done by the lower right-hand op amp through the 100 k ohm resistor connected to the motor armature. (The sensing takes place when the power MOSFET is turned off.) The upper right-hand op amp is the PWM modulator. It delivers a rectangular wave to the motor, with a duty cycle determined by the dc voltage level applied to its noninverting terminal by the previously described sense amplifier, the lower right-hand op amp. The inverting input terminal of the PWM receives a triangular wave from an oscillator comprising the two left-hand op amps. Thus, the circuit is straightforward—the PWM is an op amp fed by a variable dc voltage and a triangular wave.

Speed control occurs because of the change in the average value of armature current that can be brought about by adjusting the 1-megohm potentiometer associated with the sensing amplifier. Unlike some rheostatic speed controls, this circuit maintains the motor speed constant for any adjusted speed. Even though there are variations in applied motor voltage, in mechanical loading of the motor, or in motor temperature, this speed-control circuit will maintain the set speed at a constant value.

Efficient speed control of permanent-magnet dc motors

The motor-control circuit in Fig. 4-2 brings together three noteworthy devices that have been instrumental in simplifying, ruggedizing, and improving electronics

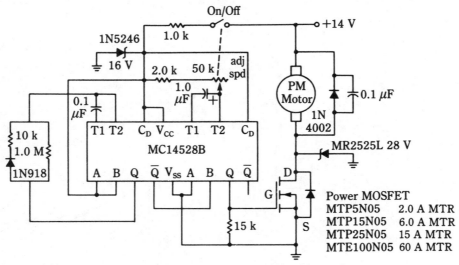

4-2 Efficient speed control of permanent-magnet motors. Motorola Semiconductor Products Inc.

control systems. These are the permanent-magnet dc motor, the power MOSFET, and CMOS (complementary metal-oxide semiconductor) logic ICs. The permanent-magnet motor, utilizing modern ceramic magnetic material has enabled the manufacture of small compact motors with much higher starting torque than was previously achieved in motors of this type. Other attributes of the motor have also been upgraded, such as speed constancy under varying load, and commutation. These motors have only two terminals; there is no wound field. To reverse rotation, it is only necessary to transpose the power lead connections to the brushes.

The power MOSFET is a very nice device for power control. Unlike the bipolar power transistor, it is immune to secondary-breakdown destruction, and does not self-destruct from thermal runaway. (That is not to say that similar catastrophic destruction could not be brought about by gross overloading and extremely abusive operation. Note, however, that the temperature coefficient of resistance of these devices is opposite that of bipolar transistors; this implies that the power MOSFET tries to limit excessive power dissipation.) And, whereas large bipolar power transistors are difficult to drive, the input impedance of the power MOSFET at dc is, for practical purposes, infinite. Finally, the switching performance of these devices is superb—it is generally easier to get efficient high-speed operation with a power MOSFET than with bipolar power transistors. Although price and relatively high ON resistance frequently loomed up as obstacles, these barriers have been largely overcome and continued developmental progress is the order of the day.

CMOS logic circuitry and linear ICs enable sophisticated circuit elements to be integrated and marketed at low cost. It is often no longer profitable, or electrically advantageous, to laboriously construct circuit functions and subsystems with numerous discrete elements and their accompanying high parts count in passive components.

In using these compelling features, the circuit of Fig. 4-2 also provides considerable design flexibility inasmuch as power MOSFETs of appropriate power capability can be selected for use with different size motors. (It is about as easy to drive a large power MOSFET as a small one.) Power MOSFETs can be readily paralleled without the power-wasting ballast resistors needed when paralleling bipolar power transistors. This is a consequence of their temperature coefficient—individual devices in a parallel-connected group do not try to hog the current.

The MC14528B is a dual monostable multivibrator. In the circuit of Fig. 4-2, one-half of this IC is connected to function in the astable mode, thereby becoming a pulse generator. The other half is deployed as a monostable multivibrator with adjustable pulse width. This adjustment varies the average current allotted to the motor, and therefore provides manual control of speed. Because the power MOSFET is either on or off, there is relatively little power lost, and overall efficiency is high, even at slow speeds. During interpulse time, the motor relinquishes energy stored in its magnetic field via the flow of current through the free-wheeling diode connected across its armature terminals. In so doing, the motor experiences more continuous torque than would otherwise be the case, and operation is smoother and more efficient because of it. However, be certain that the current capability of this diode is compatible with the size of the motor.

The combination of CMOS control logic and the power MOSFET results in very low standby power drain, a welcome feature in battery-operated systems. In most instances, it will be found that EMI and RFI is much less than for thyristor-type motor controllers.

dc motor speed control with a PWM module

Many fractional-horsepower dc motors require from 5 to 35 V and consume up to 5 A. This range falls within the capability of the Lambda LAS 6300 pulse-width modulator. Because this IC has this power capability, and because a sufficiently high frequency can be used so that the inductance of the motor armature itself serves to integrate the current through it, an extremely simple and efficient motor-speed control can be devised. No external power transistor or chokes are needed, and only a few passive components are required. The circuit of such a motor-speed control is shown in Fig. 4-3. No feedback is used; a dc voltage applied at V_C controls the speed. This voltage may be derived from a potentiometer connected across the dc voltage source, V_M. The switching rate of 25 kHz is set by the 0.025 μF capacitor connected to pin 3.

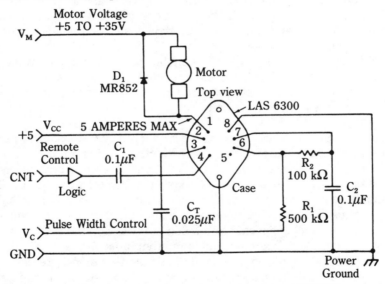

4-3 Speed control of fractional-horsepower dc motor with a PWM IC. Lambda Semiconductors

The functional block diagram of the LAS 6300 is shown in Fig. 4-4. It provides insight into the nature of the speed control, as well as some of the self-protection provisions of this IC. Pin 4 provides a useful control function. Note that in the motor-control circuit of Fig. 4-3, pin 4 is connected to a coupling capacitor and to a logic device. This setup is used in the following manner: A momentary logic 0 turns the LAS 6300 on, whereas a momentary logic 1 turns it off. Thus, the motor can be turned on and off remotely.

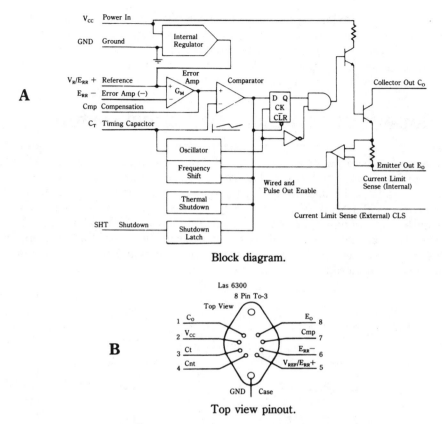

Block diagram.

Top view pinout.

4-4   Block diagram and pinout of LAS 6300 PWM control IC. Lambda Semiconductors

An interesting aspect of this motor-speed control is that excessive motor current causes the LAS 6300 to go into its current-limiting mode. This operational mode confers protection to both the IC and the motor by lowering the oscillator frequency, thereby reducing the duty cycle of the motor current. On the other hand, thermal shutdown also occurs automatically, but in this protective measure the output current pulses are totally inhibited. The experimenter should be aware, however, that thermal shutdown will not occur if the dc resistance from pin 4 to ground is less than 5000 ohms.

This is a sure-fire control circuit, but be careful with selection of free-wheeling diode D1. It must be either a Schottky diode or a fast-recovery junction diode—rectifier or garden-variety diodes will produce inefficient motor operation and will endanger the power-transistor stage in the LAS 6300. The best dc motor for speed control is the modern permanent-magnet type using ferrite or ceramic field magnets. The experimenter will also find series motors suitable for some applications. With the PM (permanent-magnet) motor, reversal of rotation is conveniently achieved by reversing the connections to the motor. With a series motor, either (but not both) field or armature connections must be reversed. A motor with a wound shunt field can also be used; the shunt field can be connected across the armature terminals. Conversely, a separate dc source for the shunt field can be provided. In any event, reversal of rotation can be brought about by simply transposing the shunt-field connections. Shunt-field current is very low relative to armature current, so this type of dc motor is easy to implement.

Unique bidirectional servo-drive system

A servo drive system involving rotational motion needs power gain, bidirectional motion, and a dead band; the latter is generally required to prevent oscillation, overshoot, or other instabilities. The conventional design approaches use linear amplifiers and special motors, such as two-phase induction types. A simpler and more economical scheme is shown in Fig. 4-5. Here, the behavior of such a servo drive system is simulated, but with a dc series motor or universal motor, and without amplifiers in the conventional sense. In order to gain insight into the operation of this circuit, first consider some clever uses of devices and circuitry elements not commonly encountered in power control.

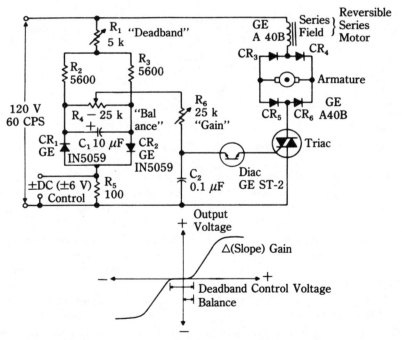

4-5 Bidirectional servo drive system. General Electric Company

In Fig. 4-6A a dc series motor is shown connected to a polarity-reversing circuit. The direction of rotation cannot be changed by reversing polarity. In order to reverse rotation, the connections to either, but not both, the armature or the field winding would have to be transposed. However, by surrounding the motor with

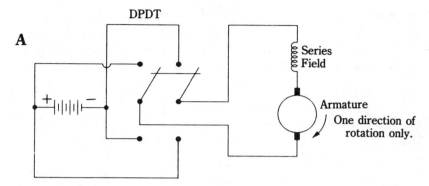

Reversing polarity to series motor does not reverse direction of rotation.

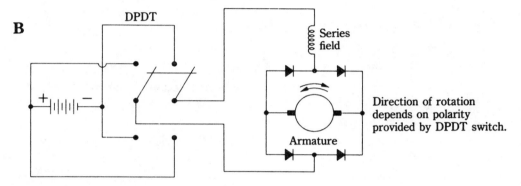

With steering-diode circuit, polarity reversal changes direction of rotation.

4-6 How the rotational direction of a series motor can be selected.

steering diodes as depicted in Fig. 4-6B, such transposition is not necessary; in this case, reversal of applied polarity, as provided by the dpdt (double-pole double-throw) switch, suffices to reverse rotation. It is also interesting to note that if the battery or dc source were replaced by an ac source, the motor in Fig. 4-6A would run, whereas the motor in Fig. 4-6B would not run—its average torque would be zero. Thus, by appropriate use of the steering diodes, some of the natural characteristics of the series motor can be electronically modified.

Another unusual deployment of components is illustrated in Fig. 4-7 which shows the triggering characteristics of diac-triac combinations under usual and unusual conditions. In Fig. 4-7A, the waveforms of a diac and triac in conventional phase-control circuits, such as light dimmers, are shown. The salient feature is the ac symmetry of these waveforms. A phase angle of approximately 90 degrees is represented in the drawings; but no matter what the phase angle, the triac waveform remains symmetrical, i.e., it is always a pure ac wave with no dc component. In the situation depicted in Fig. 4-7B, a dc bias is applied to the diac so that voltage breakdown occurs for only one polarity of the ac wave. As a consequence, the

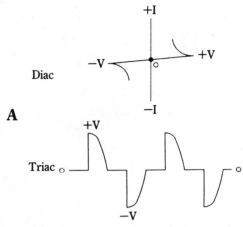

Diac

A

Triac

Waveforms in normal diac-triac control circuits.

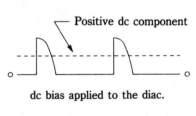

4-7   Effect of dc bias on the diac.

B

Positive dc component

dc bias applied to the diac.

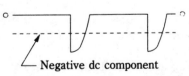

C

Negative dc component

Reversed polarity of dc voltage applied to diac.

triggered waveform of the triac is asymmetrical and has an obvious dc component. If the dc bias applied to the diac is reversed in polarity, the situation is then as shown in Fig. 4-7C. Thus, a low-power dc signal can control the dc polarity of relatively high-powered dc output from the triac. In a sense, the triac behaves somewhat like an SCR when used in this manner in that SCR action provides rectification, but conventional use of the triac does not.

By combining these unique implementations of the motor and triac, the operation of the circuit of Fig. 4-5 becomes straightforward. Essentially, it can be viewed as combining an electronically modified motor with a polarity-selectable high-gain dc source. Although it is not commonplace to think of a thyristor, such as a triac, as providing power amplification, it is only because such a viewpoint is not particularly convenient in most applications. In this particular case, the concept of power amplification is meaningful inasmuch as a low-level variable-polarity dc signal controls the activity of a powerful motor in the same fashion as is ordinarily done with a linear or analog amplifier.

The allusion to a reversible series motor in Fig. 4-6 implies that this is the optimum type of motor for the purpose; actually, all dc series motors, as well as all universal motors, will work in this setup. If, however, a series motor is not designated as reversible, it often happens that brush alignment is off geometric center in order to favor commutation for one direction. If such a motor has adjustable brushes, it is generally best to reposition them. If not, it may be found that performance is satisfactory nevertheless.

In most servo systems using series motors, it has been necessary to disconnect the internal connections between field windings and the armature. This surgery is not always trivial, especially in universal motors where the manufacturer assumed these connections would be permanent. Obviously, a compelling feature of reversible series motors is that such motors can be used in their natural state. Incidentally, there appears to be little reason why the competent experimenter couldn't adapt features of this scheme for other types of dc motors. With a permanent-magnet motor, the steering diodes would not be needed. With a shunt-field motor, it probably would be necessary to insert a rudimentary filter between the triac and the field winding.

This servo driver can readily be converted into a full-fledged servo system by the addition of an appropriate negative-feedback loop and a reference voltage. A straightforward implementation would involve a small dc tachometer mechanically coupled to the motor, and a resistive network for combining the tachometer, reference, and control voltages at the input of the system.

Full-wave phase control of motor with an SCR

An interesting phase-control circuit for universal and dc motors is shown in Fig. 4-8. Here, full-wave power is applied to the motor even though only a single SCR is used. The basic principle underlying this implementation can be understood from a contemplation of the simplified circuit of Fig. 4-9. This circuit shows that a bridge rectifier appropriately associated with the single SCR enables the overall circuit to apply phase-controlled ac to the load. The waveforms will be seen to be essentially identical to those ordinarily provided by antiparallel connected SCRs,

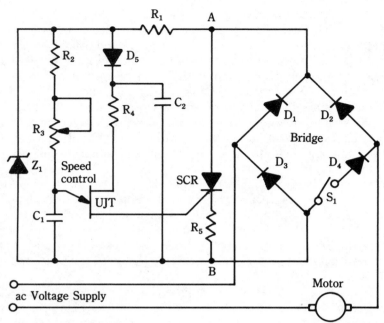

SCR Motorola MCR 808-4

UJT MU970

D1,D2 1N3493 — 200 V, 18A Rectifier

D3,D4 1N3493R — 200 V, 18A Reverse Polarity
       Rectifier

D5   1N4001 — 50 V, 1A Rectifier

Z1 1N751A — 5.1 V, 400mW Zener Diode
R1 18 k, 2W
R2 3.9 k, 1/2W
R3 50 k, Potentiometer
R4 330Ω 1/2W
R5 See table
C1 .1μfd
C2 10μfd, 10 V

Nominal R5 Values

Motor Rating (A)	R5 (Ω)	(W)	
2	1	5	R% = 2 / IM
3	0.67	10	
6.5	0.32	15	IM = Max Rated Motor Current (RMS)

4-8 Full-wave motor control circuit using a single SCR. Motorola Semiconductor Products, Inc.

as well as to the more familiar triacs. Possible justifications for utilizing this basic
arrangement rather than a triac could entail one or more of the following factors:

- Both SCRs and rectifying diodes are capable of greater electrical ruggedness than the triac. Both of these solid-state devices can be readily obtained
  in higher voltage, current, and power ratings than triacs.

- Both SCRs and rectifying diodes can be obtained with greater frequency capability than triacs.

- Comparing cost on a dollar per watt, or dollar per peak ampere basis, some applications would merit consideration of the SCR plus rectifier bridge approach.

The objective of providing full-wave rather than half-wave power to the motor is to obtain smoother performance, wider control range, and greater maximum torque or speed. Also, if speed regulation is involved, the full-wave circuit generally provides better stabilization. The availability of bridge-rectifier modules enhances the convenience of constructing such a circuit.

In the circuit of Fig. 4-8, the firing of the unijunction transistor is advanced more as increased motor current flows through resistor R_5. Thus, when the motor is loaded and tends to lose speed, it generates less counter EMF and the manifestation of this is greater motor current. Greater motor current develops a higher voltage drop across R_5; the polarity of this voltage drop is such that the firing of the UJT (unijunction transistor) is attained sooner than the time that was initially set in by adjustment of R_3. In other words, as the motor experiences more mechanical loading, it takes less time for capacitor C_1 to charge to the firing potential of the unijunction transistor. The result of this timing advance in the UJT relaxation oscillator is earlier per-cycle triggering of the SCR, and therefore increased motor voltage. This, in turn, is accompanied by increased motor torque, which is evidenced by higher speed. Thus, the behavior of the system is such that a nearly constant speed characteristic is maintained with respect to motor loading; the corrective tendency works both ways—a relaxation in motor loading will also invoke action preventing any appreciable increase in speed.

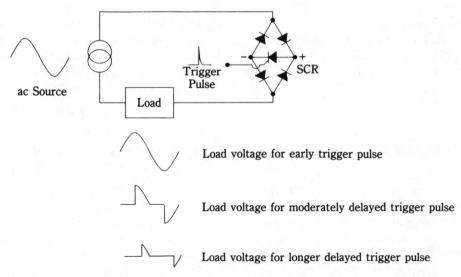

4-9 Simplified circuit showing how full-wave control is obtained from one SCR.

Explanations are in order regarding some of the details of this circuit. Zener diode, Z_1, in conjunction with R_1, forms a dc source for the operation of the UJT relaxation oscillator. However, after the triggering of the SCR, the available voltage at points A-B is less than the breakdown voltage of the zener diode. This would tend to make the UJT vulnerable to undesirable firing—lowering of the interbase voltage in these devices reduces the required firing potential. To safeguard against this, a memory circuit is formed of diode D_5 and capacitor C_2. Here, C_2 retains the approximate voltage developed across Z_1 prior to the drop of voltage AB. Inasmuch as C_2 is a large capacitor, it takes over as dc source for the interbase circuit of the UJT during conduction of the SCR.

Nothing has been said about commutation of the SCR. Actually, no special provision is needed; conduction ceases in the SCR at the zero crossing of the ac voltage wave. That is why there is no filter capacitor across points A and B—such filtering would produce good direct current for the motor, but the circuit would hang up and be inoperative. In order for the circuit to work, the SCR must automatically turn off at the end of each half cycle. Fortunately, series and universal motors can be reasonably happy on ac waveforms.

Switch S_1, in the rectifier-bridge arm, provides the option of half-wave operation. Under most circumstances, superior motor operation is obtained from the full-wave format. However, under some load conditions, it might happen that instabilities such as hunting, or erratic speed changes, may be experienced; changing to half-wave operation can sometimes beneficially alter the feedback parameters and alleviate such poor performance. Another use for half-wave operation is with shunt, or permanent-magnet dc motors. Such motors will operate with their armatures impressed with pulsating dc or unidirectional current, but not with ac waveforms.

It is not necessary to duplicate the designated semiconductor devices of this circuit; the components, with the exception of R_5, are not critical. Much will depend upon the type of motor used and the load variations it is subjected to. Although the table associated with the parts list provides nominal values for R_5, it would be a good idea to use an adjustable wirewound type of resistance, or a rheostat. In any event, this component should be of good quality and conservative rating. Changes in resistance because of temperature coefficient or faulty contacts are likely to adversely affect speed regulation.

ac and dc motor control with a power op amp

The RCA HC2500 is a hybrid operational amplifier with considerable power capability; it can provide 100 W of RMS power to a load, and can deliver 7 A of peak current. The total dc supply voltage can be as high as 75 V and either single or split dc supplies can be accommodated. Such ratings make this device useful for control of small motors. Implementation is simple inasmuch as no external power transistors are needed, as would be the case with conventional op amps.

Two experimental motor-control applications are shown in Fig. 4-10. The setup depicted in Fig. 4-10A allows the speed of an induction motor or synchro-

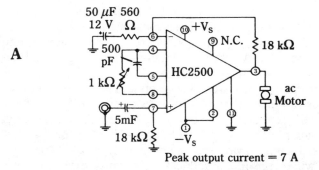

Peak output current = 7 A

Controlling speed of induction motor by frequency variation.

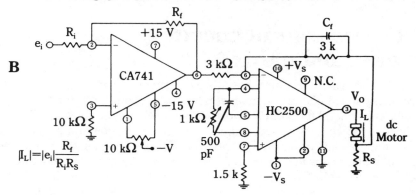

Controlling and regulating current in a dc motor.

4-10 Motor-control circuits using the HC2500 power op amp. RCA

nous motor to be controlled via frequency variation. The idea is to feed the input of the amplifier from a small variable-frequency source. In essence, the power op amp then functions as an inverter, converting the energy from the dc power supply to alternating current. This is a very efficient means of controlling the speed of these motors. (Very little speed change can be brought about by varying the voltage or current applied to such motors.) Note that the so-called universal motor, although commonly powered by ac, would not undergo speed variation in this circuit—in such motors, speed is essentially constant with respect to frequency. (Also, there would be no point in using the universal motor in this arrangement because this type of motor will run on dc.)

Two things are accomplished by the scheme in Fig. 4-10B. The current to a dc motor is regulated so that it remains constant during load variations. This is considered desirable for various applications of powered tools; both the motor and the controlling amplifier are protected by this operational mode. Secondly, the actual level of the regulated current can be manually controlled by the dc voltage, e_i, applied to the input of the system. Inasmuch as the torque developed by a dc motor is a function of its current, this motor-control technique is best described as torque control and regulation. In practice, it will be found that the motor must readjust its speed to meet a given torque command. Therefore, such a control method may serve as a speed control for certain purposes. There will be no speed regulation, however. Indeed, speed will change more as a function of load variation than would be the case without current (torque) stabilization.

Permanent-magnet motors work well in this setup. The series motor will work too, but it will be deprived of its natural ability to meet high torque demand by drawing high current. Current-sensing resistor R_S should be as high as possible without degrading motor performance. It will generally be in the fractional-ohm region. C_f is also experimental; it protects the op amp from the consequences of abrupt load changes where the motor inductance permits development of high voltage. Essentially, this capacitor limits the transient response of the amplifier.

SCR speed-control circuits for series and universal motors

For a number of years, technical literature has abounded with simple SCR circuits for providing speed control of small series-type and universal motors. These circuits take the general form of the two speed-control circuits shown in Fig. 4-11. At first glance, the circuits appear similar enough to provide similar performance. This is not quite true, however. The circuit in Fig. 4-11A is similar to popular light dimmer circuits, and would serve such a purpose if a lamp were substituted for the motor. However, substituting a lamp for the motor in the circuit of Fig. 4-11B would not produce the same circuitry action as the motor. Although there is no obvious feedback path in circuit B, speed regulation occurs with the motor in the circuit. If the two circuits were compared, the loss in motor speed would be much less in circuit B than in circuit A when equal mechanical loads are imparted to the motors, assuming the important motor-operating parameters were initially the same. In both circuits, a heat sink should be provided for the SCR.

Because of the nature of an SCR, the motor in circuit B can receive current for no longer than a half cycle, or 180 degrees. Because of the phase-control provision, the duration of the current pulse delivered by the SCR will generally be less than this. However, while the SCR is blocking the passage of motor current, the motor is turning and is therefore simultaneously behaving as a generator. Circuit B is such that the counter EMF the spinning motor develops because of its generator action opposes the gate-cathode voltage of the SCR. If the added load tends to slow the motor down, this self-generated voltage will decrease, allowing the SCR to trigger earlier in the ac cycle, thereby delivering more current to the motor than would otherwise have been the case. The added current manifests itself as electromagnetic torque, tending to speed the motor up. The closeness of the regulation depends considerably on the motor, but the fact that corrective feedback is involved is evident.

These circuits were designed for 35 V dc series motors with $1/15$ horsepower ratings. The circuits can be scaled up and down to accommodate other motors. In

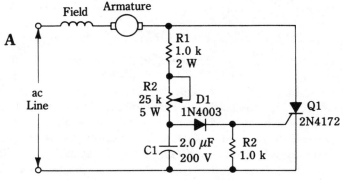

Speed control without regulation.

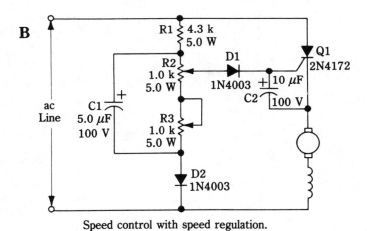

Speed control with speed regulation.

4-11 Half-wave phase-control circuits for series and universal motors.
Motorola Semiconductor Products, Inc.

any event, it makes sense to know what kind of performance to expect. For some applications, the inherently poor speed regulation of the series motor is suitable, or even desirable. In other applications, a more nearly constant speed characteristic is best. In the latter case, the invisible feedback path in circuit B electronically modifies the natural behavior of the series or universal-type motor.

When either circuit is used with a universal motor, it sometimes proves useful to connect an spst (single-pole single-throw) switch across the cathode-anode terminals of the SCR. Then, extra performance is forthcoming from the motor when the switch is closed. This jump is not as abrupt as it might initially appear; if the SCR is conducting very close to a half cycle, the closure of this switch will increase RMS motor current by approximately 40 percent. The control can be made idiot-proof by mechanically linking the switch to the phase-control potentiometer.

Regulation and speed control with a simple circuit

A simple circuit for imparting both speed control and speed regulation to small universal-type motors is shown in Fig. 4-12. Speed variation is produced by changing the triggering time of the SCR, causing the motor to receive varying portions of half cycles of the applied ac. This is somewhat reminiscent of phase-controlled light dimmers, but it will be noted that there is no phase-shift network. Even more intriguing is the lack of any obvious feedback path. (Regulators generally have sensor circuits, error amplifiers, and well-defined feedback paths.) It will be interesting to see how this control and regulating scheme operates, for its simplistic format belies its capabilities.

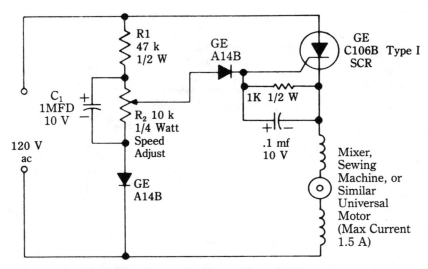

4-12 SCR speed control with speed regulator. General Electric Company

First, consider the nature of the universal motor; it is essentially a series motor, differing only in minor details from true dc series motors. When this motor is rotating, it generates a counter EMF. In the circuit, it not only generates the counter EMF when the half-wave pulses from the SCR are applied to it, but it continues to generate this counter EMF during the nonconducting half cycle of the SCR. Secondly, consider the triggering characteristic of the SCR—the greater its anode-cathode voltage at a given time, the less gate voltage it needs for triggering. Combining these phenomena leads to the explanation of actual circuit operation.

Assume the motor is running at some speed determined by the setting of potentiometer R2. If, for some reason, the mechanical load on the motor is increased, the motor will try to slow down; this is the natural operating behavior of series and universal motors. By slowing down, the counter EMF decreases, allowing the motor to consume more current and thereby develop the higher torque needed to accommodate the greater load. However, in this circuit, when the motor slows down the reduced counter EMF results in a higher net voltage across the anode-cathode terminals of the SCR. This is so because this net voltage is determined by the balance between the voltage impressed from the power line, and the voltage from the generator action of the motor (the counter EMF). The SCR is now able to fire with less gate voltage than it required before the motor slowed down. This is tantamount to saying that the SCR triggers earlier in the ac cycle, thereby delivering increased voltage to the motor, speeding it up again. The overall action is one of regulation in which the motor tendency to drop speed is largely corrected. The opposite sequence of events occurs if the motor tries to gain speed. The net result is that the motor, which by itself has very sloppy speed behavior, is electronically converted to a constant-speed machine.

The components designated in Fig. 4-12 are from an early design. Other parts and devices can be substituted, but best low-speed control will result from the use of a sensitive-gate SCR. This type of trigger control is limited to an effective phase delay of 90 degrees, so there is inherently less low-speed control than in SCR circuits using RC phase-shift networks, as in light dimmers, which are also used for motor control (but without regulation).

ac tachometer function via special optocouplers

The speed of a dc motor may be nicely controlled by a phase-locked loop. In such a system, the feedback signal must be ac and its frequency must be proportional to motor speed. The idea of using a slotted disk in conjunction with an optical-electronic arrangement is an old one. However, the implementation of such a scheme has generally involved a few headaches because of the mechanical details that had to be taken into account. In order to circumvent this, TRW and other firms have made available a family of optocouplers, so fashioned as to facilitate such applications.

Figure 4-13 depicts a simplified phase-locked loop system for controlling the speed of dc motors. The permanent-magnet motor is frequently used. In some applications, a shunt-wound dc motor offers the advantage that the shunt field requires much less power than does the armature, or the entire motor. If, however, control is imparted through the shunt field, it must be kept in mind that the motor speed decreases with increasing field current and the design must be altered accordingly. In any event, the salient feature of the special optocouplers is their physical configuration; they are two-prong forks with an optical source in one side, and a detector on the other side. The slotted disk rotates between the optical source and detector, thereby producing optical chopping. The result of this is an output of pulses with a rate proportional to speed and the number of slots in the disk.

The optical source is usually an LED or an infrared-emitting diode. The detector can be a photoresponsive semiconductor device, such as a photodiode, a phototransistor, a photo Darlington, or a photo FET. Other devices, such as filamentary lamps and photoconductive cells have been used in the past, but they tend to introduce

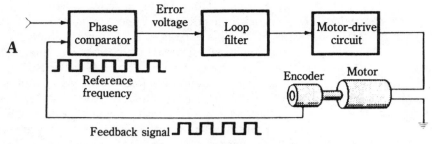

A

Simple phase-locked loop system.

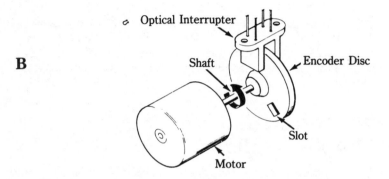

B

Implementation of the special optocoupler, known as a slotted optical switch.

4-13 Use of an optocoupler in a phase-locked loop motor-control system.

problems with reliability, temperature dependence, and frequency response. Other schemes can produce analogous results; for example, magnetic reluctance and magnetic induction versions of the optical chopper are feasible too. The new optocoupler devices offer the advantage that fine speed control resolution is easily realized by merely introducing a number of slots on the disk. This can be done in conjunction with the insertion of a programmable frequency divider in the feedback loop.

These optocoupler devices are known as slotted optical switches. Figure 4-14 depicts one of the many such slotted optical switches made by TRW. This, the OPB837, features convenient mounting tabs, and utilizes a photodiode as infrared light detector. As can be seen, it is only necessary to insert an encoder disk in the slot to produce ac tachometer action (that is, to obtain a pulse train with a repetition rate proportional to speed). The disk can have a single hole for basic operation, or it can have many holes in order to produce high resolution in an appropriate system design. Other detector schemes used in TRW slotted optical switches are shown in Fig. 4-15. Circuits in Fig. 4-15C and D contain linear amplifiers and Schmitt triggers; these circuits are available to provide either buffer, or inverter action. The circuit in Fig. 4-15B can be used to provide a unique function, which is described below.

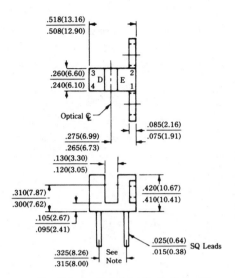

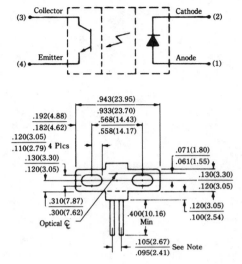

Note: Dimensions Controlled at Housing Surface Only.

Dimensions Are in Inches (Millimeters).

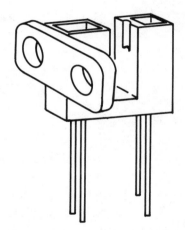

4-14 A typical slotted optical switch. TRW Inc.

The circuit of Fig. 4-15B comprises a pair of infrared-emitting diodes and their phototransistor detectors. Thus, there are two aperture systems in the fork of the optocoupler device. In the TRW OPB822 family of optical interrupters, the aperture pairs are arranged side by side. This can be clearly seen in the drawings of Fig. 4-16. The basic idea of such an arrangement is that the chopped signals produced by the dual detectors can be used with appropriate logic circuitry to indicate the direction of rotation of the encoder disk. The logic circuit for accomplishing this is shown in Fig. 4-17.

The outputs of the two detectors can be considered as channels A and B in conformity with Fig. 4-17. The output of channel A is voltage amplified by transistor Q1, and imparted TTL compatible rise and fall times by the Schmitt trigger designated L1. Similar action occurs in channel B via Q2 and L2. Observe that the

A

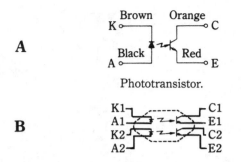

Phototransistor.

B

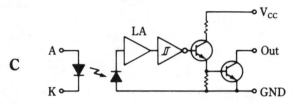

Dual-channel phototransistor.

C

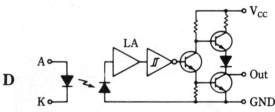

Open-collector logic-compatible output.

4-15 Additional formats in the slotted optical switch family. TRW Inc.

D

Totem pole logic-compatible output.

channel-A Schmitt trigger clocks the latch and the channel-B Schmitt trigger connects to the D input of the latch. With this setup, the logic state of the latch Q output tells the direction of rotation of the encoder disk. Now, see how this comes about.

Suppose the logic state of the Q output of the SN7474 is high. For this to be so, channel B had to be turned on prior to the turn on of channel A. In other words, when the edge-triggered latch was clocked, there already was a logic-high signal at its D input. Under such conditions, the truth table of the latch would show that its Q output would go high. Referring to Fig. 4-16C, counterclockwise rotation exists for the condition that channel B turns on prior to channel A. The opposite conditions prevail for clockwise rotation and the Q output of the latch is then low. As shown in Fig. 4-17, an LED indicates rotational direction, but the output from Q of the latch could be used to initiate some function in the mechanical system associated with the motor.

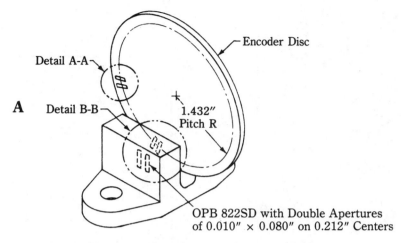

Setup with encoder disk showing double aperture system.

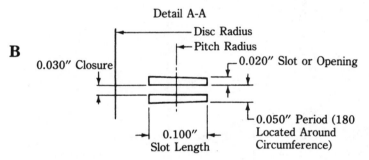

Optimum slot dimensions and spacing in encoder disk.

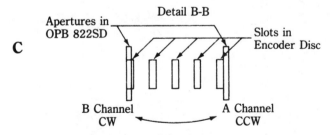

Relationship of slots to aperture.

4-16 Two-channel interruptor for determining direction of rotation. TRW Inc.

Logic-actuated switch provides isolation

Photocouplers or photon-coupled isolators are commonly thought of as associations of light-emitting devices and photodiodes or phototransistors as detectors.

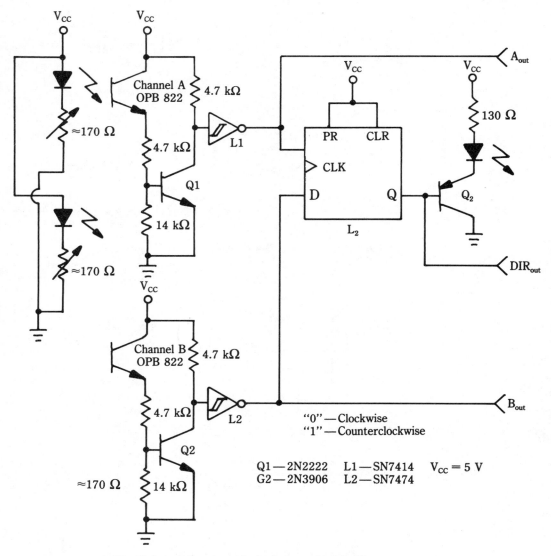

4-17 Circuitry for use with dual-channel optical interruptor. TRW Inc.

Unique components incorporating photon isolation, but utilizing a light-controlled thyristor are made by General Electric Company. The thyristor acts as a gateless triac, or as a diac, in the sense that it provides bilateral conduction when turned on by the ac applied to it through a load circuit. This characteristic makes it useful as a trigger source for an external triac, which might, in turn, act as a switch for a large load, such as an induction motor.

Just such an application is shown in Fig. 4-18. With a small parts count, the features of electrical isolation and receptivity to logic-level control are achieved.

The basic purpose for using this scheme would be to avoid the contact and mechanical problems encountered when using electromagnetic devices (contactors and relays) for switching multiampere inductive loads. The circuit is also useful for turning on and off resistive loads such as heaters and lights. In all cases, consideration should be allowed for inrush currents—stationary motors, inactive heaters, and cold filaments in lamps all consume surge currents when first turned on. That is why a fairly large triac, the SC160D, is indicated in Fig. 4-18.

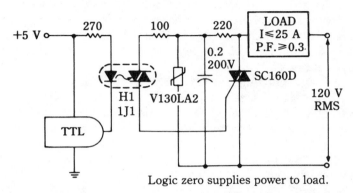

4-18   Logic-actuated contactor suitable for ac induction motors. General Electric Company

The V130LA2 is a metal-oxide varistor. This General Electric device absorbs the energy of voltage transients. It protects the thyristor element in the J11J1 and prevents accidental turn on of the triac from line transients. It also acts to reduce electromagnetic interference when the triac is actuated. A practical hint involves the connections of unused pins on the 6-pin dual in-line package of the H11J1. Whereas pin 3 is truly a blank pin, pin 5 is internally connected to the substrate of the photothyristor. Because of this, pin 5 should be left floating—it must not be grounded or either purposely or inadvertently connected to any part of the external circuit. Otherwise, the photothyristor will not be properly responsive to light actuation.

Electronic switch for starting single-phase induction motors

The single-phase induction motor is a wonderful machine in that it is inherently rugged and is easy to manufacture. It has no brushes, commutators, slip rings, or other sliding contacts. Its electrical and mechanical characteristics are about what the doctor ordered for a wide variety of tasks, especially around the home. But unfortunately, it develops no starting torque in its basic form—something has to be done to initially get it going. Once brought up near its normal running speed, the starting mechanism, whatever its nature, can be dispensed with. Normal running speed is something close to synchronous speed. For the commonly encountered 60 Hz four-pole types, this is slightly less than 1800 rpm (revolutions per minute). In 60 Hz two-pole induction motors, this is slightly under 3600 rpm.

A popular means of making these motors self-starting is to design them with an extra stator winding and to impress this starting winding with a current 90 degrees out of phase with that applied to the main or running-stator winding. The phase displacement is produced by a large electrolytic capacitor connected in series with the starting winding. Then, after the motor has accelerated up to perhaps 80 percent or so of its normal running speed, a centrifugal switch disconnects the starting winding and the motor is able to rely on its own torque to bring it to running speed. As long as the starting winding, with its capacitor, is in the circuit, the motor behaves essentially as a two-phase machine. (The two-phase induction motor develops a rotating field, and therefore is self-starting. Three-phase motors, for the same reason, develop high torque at standstill, and are self-starting without any switching mechanism.)

The trouble with centrifugal switches and electrolytic capacitors is that they are both relatively high maintenance items. When the single-phase induction motor gives trouble, the source of the poor performance is rarely in the motor itself, unless conditions have become sufficiently abusive to burn out the stator winding(s).

The electronic-switching scheme shown in the motor-start circuit of Fig. 4-19 provides a partial solution to this dilemma. Although it retains the electrolytic capacitor, it eliminates the centrifugal switch. This remedy is a good one because it is much easier to correct for a bad capacitor than it is to overhaul or replace a defunct centrifugal switch. The operation of this circuit is predicated upon the inrush current of a motor starting up and gathering speed. At first the inrush current is high and sufficient voltage drop is developed across R1 to keep the triac triggered. This, in turn, keeps the phase-shifting capacitor, C2, in the starting-winding circuit. When the motor has attained sufficient speed to be on its own, the decreased current through R1 will no longer develop sufficient voltage drop to trigger the triac, thereby opening the starting-winding circuit. Obviously a bit of experimentation is in order for the optimum value of R1. If this resistance is too high, the starting circuit will not be opened during normal running and both the motor and the triac will overheat.

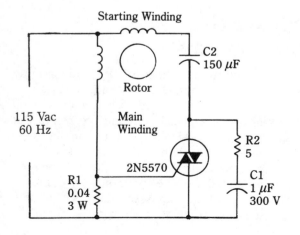

4-19   Electronic switch for capacitor-start inductor motors. Motorola Semiconductor Products, Inc.

Triac phase-control circuit for speed adjustment

A triac tends to be a more satisfactory device for speed control of universal motors than ordinary SCRs. Although both devices use phase control to vary the current through the motor, the triac, being a full-wave device, symmetrically controls the phase of both half cycles of the applied ac. The resultant full-wave current format then produces smoother motor operation than can readily be attained from the half-wave rectification of SCRs. This tends to be particularly noticeable at low speeds. The triac phase-control circuit shown in Fig. 4-20 is depicted for a family of 12 A triacs. However, the same component values generally suffice for both smaller and larger devices as well.

A generalized load is indicated in Fig. 4-20 rather than a motor symbol because this circuit is also widely used for light dimmers and for the control of

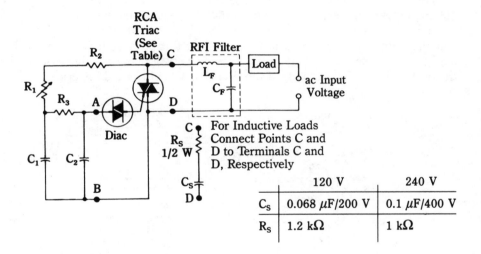

	120 V	240 V
C_S	0.068 μF/200 V	0.1 μF/400 V
R_S	1.2 kΩ	1 kΩ

ac Input Voltage	C_1	C_2	R_1	R_2	R_3	RFI Filter		RCA Types
						L_F* (typ.)	C_F* (typ.)	
120 V 60 Hz	0.1 μF 200 V	0.1 μF 100 V	100 kΩ ½ W	2.2 kΩ ½ W	15 kΩ ½ W	100 μH	0.1 μF 200 V	BTA23B,C
240 V 50 Hz	0.1 μF 400 V	0.1 μF 100 V	250 kΩ 1 W	3.3 kΩ ½ W	15 kΩ ½ W	200 μH	0.1 μF 400 V	BTA23D,E
240 V 60 Hz	0.1 μF 400 V	0.1 μF 100 V	200 kΩ 1 W	3.3 kΩ ½ W	15 kΩ ½ W	200 μH	0.1 μF 400 V	BTA23D,E

*Typical values for lamp-dimming circuits.   92CS-33761

4-20 Triac phase-control circuit for speed adjustment of universal motors. RCA heaters. Often, a dc series motor will also work, but efficiency and commutation will not be as good as with the universal motor.

This circuit features a double time constant phase-shift network. As with SCRs, this reduces hysteresis in the firing of the triac, thereby making the manual adjustment of speed more repeatable. (This control technique provides for adjustment of motor speed, but does not produce speed regulation.)

Adding to the reliability of this circuit is the diac trigger device. This is essentially a gateless thyristor designed to break down on both polarities of the ac wave. A given diac can properly trigger a wide variety of thyristors. The elemental low-pass filter comprising L_F and C_F attenuates much of the radio-frequency interference that would otherwise get back to the power line. Such high-frequency energy is generated by the extremely rapid turn-on time of the triac. Radiation from the power line can cause much trouble.

From the table in the schematic diagram, note that this 12 A family of triacs includes devices rated for 240 V as well as 120 V operation. These are, however, intended for 50 and 60 Hz service. Special triacs are marketed for operation from 400 Hz power sources; triacs for even higher frequencies have been specified for special applications, but are not readily available. The 400 Hz phase-controlled circuit shown in Fig. 4-21 uses triacs specified for this power-line frequency. This circuit uses different RC values in the phasing network, but is otherwise similar to the 60 Hz circuit. The triac is of the T41113, T41114, T41115 series.

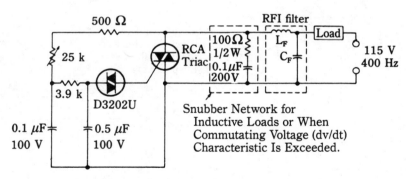

4-21   400 Hz phase-controlled triac circuit. ʀᴄᴀ

Transistors as motor-control devices

A pause is in order to contemplate the status of transistors power-control devices for motors. Extensive applications were visualized when the power transistor made its debut. However, early transistors did not survive the harsh nature of motor loads too well.

Motor-control techniques have generally been dominated by thyristor power devices, notably the SCR and the triac. There have been exceptions to this, depending upon the size of the motor and whether it was an ac or dc type. However, for motor sizes beyond about the one-tenth horsepower rating, bipolar transistors have been at a disadvantage. They tended to lack the voltage and current ratings required, and were further handicapped by their one-polarity operational requirement. Of course, various stratagems can be used to tolerate these short-comings, such as paralleling, the use of diodes, and novel circuit arrangements. The modern trend still makes use of transistor power devices, but they now assume the following more useful formats:

- Bipolar transistors are now available with much better ratings—these include current, voltage, SOA, and thermal resistance. Garden-variety power transistors of the past are overshadowed by the newer versions for motor control.

- The bipolar Darlington transistor, for many years a lackluster power device, has become a capable brute-force element with potential for controlling both fractional- and integral-horsepower electric motors. Further facilitating practical application, the Darlington device is generally easier to parallel than single-transistor types. These rugged Darlingtons still retain their easy-drive characteristics because of their relatively high beta ratings.

- Both the modern single-element bipolar transistor and the newer Darlingtons are available in PNP as well as NPN polarities. Often, there is a very close match between NPN and PNP counterparts. This simplifies the design and implementation of push-pull and bridge power amplifiers; it also helps provide for regenerative or dynamic braking and allows for more straightforward design of polyphase circuits. Complementary-symmetry power-amplifier circuitry making use of NPN-PNP devices usually features a relatively low parts count. Thus, the traditional concept that the PNP format was unsuitable for high-power devices is no longer valid.

- Power MOSFET devices have progressed considerably since their commercial debut. At first a novelty, these devices now successfully compete with conventional bipolar transistors, with power Darlingtons, and with thyristors. They have intrinsically fast operating characteristics, and tend to be immune to thermal runaway. For certain motor-control applications, these devices possess the unique feature that they can operate with either polarity. Even more design flexibility results from the fact that both N-channel and P-channel types are available. (This corresponds respectively to NPN and PNP bipolar types.)

Motor control with a dual-power op amp

The L272 is a dual op amp with power capability. Whereas commonly used op amps, such as the popular 741, generally cannot safely provide more than 100 mA, the output-current capability of each of the L272 op amps is 1 A. Because these devices can also operate from a 24 V source, applications to the control of small motors are feasible. Such control is facilitated by the differential input and the push-pull output configurations. As with other ICs, internal circuitry is direct-coupled. This enables easy application to permanent magnet, series, shunt, and other dc motors. However, series dc motors and universal motors will not reverse their direction in the illustrated circuits. (In such motors, direction of rotation is not changed by polarity reversal at the terminals. To change direction, either the series-field connections, or the armature connections, but not both, must be reversed.)

The circuit shown in Fig. 4-22A enables control of direction and speed by means of an analog control voltage. This neatly circumvents the need for a dual power supply which would be needed if only a single op amp were used. A nice feature of this arrangement is that the control voltage for zero speed (standstill) tends to be the same regardless of the main supply voltage. Thus, for many purposes, the system can be operated from a simple, nonregulated power supply.

The circuit arrangement of Fig. 4-22B provides for logic-controlled direction of rotation. This is a good peripheral output interface for certain microprocessor systems where it is desired to select any of three motor-operating conditions—clockwise rotation, or counterclockwise rotation, or standstill. Standstill occurs when both input terminals are impressed with like logic levels. Unlike logic signals at the inputs cause rotation in one direction or the other. Thus, the arrangement functions like a single-pole double-throw switch from the viewpoint of the motor. In order for this scheme to be operative, the main supply voltage, V_{S2}, must be

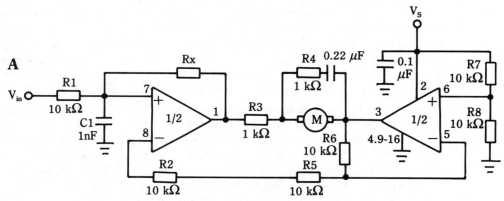

Speed control from a variable voltage source.

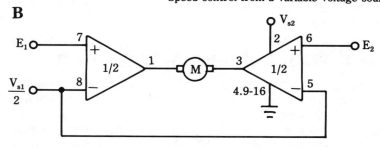

Direction selection from a microprocessor or other digital source.

4-22 Bidirectional motor control with the L272 power op amp. SGS ATES Group greater than the logic supply voltage, V_{S1}. Because the logic supply voltage will usually be 5 V, this requirement is readily met in practice. Note that the logic voltage levels are applied to terminals 6 and 7. The $V_{S1}/2$ voltage applied to terminal 8 sets the standstill condition of the motor.

Full-wave motor controller using antiparallel SCRs

The circuit of Fig. 4-23 can be said to be a classic approach to full-wave phase control of a motor. The implication is that the simplest devices are used, these being SCRs, transistors, and diodes—there are no triacs, diacs, or unijunction transistors. And of the devices used, only the SCRs carry the motor current. Some designers opt for this approach on grounds that you can anticipate greater reliability, especially where abusive overload conditions are encountered. In some measure, such a contention is subjective, evidencing the biases of the designer. To give credit where due, however, the SCR tends to be a more rugged device than the triac. If need be, you can produce SCRs with voltage, current, frequency, and peak-current ratings far beyond what is available in triacs. Also, in some instances, commutation problems in triacs are less likely to find counterpart poor performance with SCRs. Perhaps the circuit of Fig. 4-23 would not appeal to the parts-count conscious constructor; yet, the circuit would hardly qualify as a complicated one.

The SCRs are connected in an antiparallel mode, or back-to-back. This mode enables the pair to simulate the operation of a triac; indeed, explanations of triac behavior often make use of the back-to-back SCR analogy. The overall result of such an analogous circuit is reasonably close. A practical difference is that the triac enjoys the luxury of simple triggering via a single gate, whereas the SCR duo requires individual triggering of the separate gates. This is remedied easily enough by the use of a pulse transformer with two secondaries. With this provision, the primary of the transformer looks like the gate of the triac. It will be noted that the secondary windings of the pulse transformer are phased to forward bias both gates simultaneously. However, only that SCR in which the anode is positive relative to the cathode fires. This implies that sequential trigger pulses produce alternate firing of the SCRs. A little thought experimentation with this principle will reveal that a single secondary winding would not work, nor would a center-tapped winding be feasible—two conductively isolated windings are at the heart of this scheme.

Another interesting circuitry adaptation is found in the trigger generator comprising complementary bipolar transistors. This exact circuit is found in the technical literature dealing with the analogous circuit of the SCR; it is also used in explaining such negative-resistance devices such as the SGS, the PUT (programmable unijunction transistor), the diac, and the UJT. The salient feature of this circuit is that it is regenerative—the collector of each transistor feeds the base of the other.

Suppose that both transistors are initially in their nonconductive state. If a

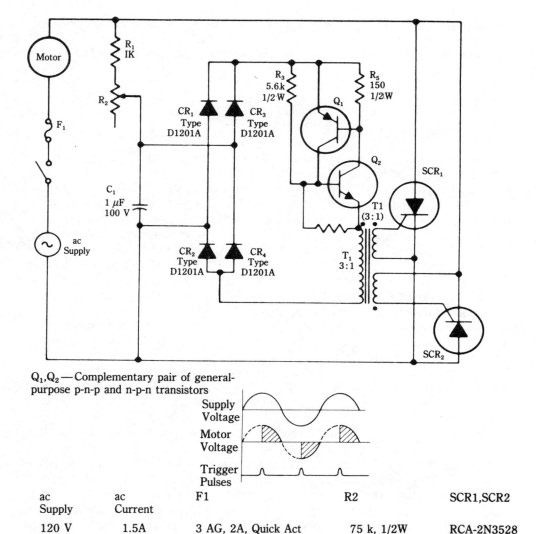

Q_1, Q_2—Complementary pair of general-purpose p-n-p and n-p-n transistors

ac Supply	ac Current	F1	R2	SCR1,SCR2
120 V	1.5A	3 AG, 2A, Quick Act	75 k, 1/2W	RCA-2N3528
120 V	5A	3 AB, 5A	75 k, 1/2W	RCA-2N3228
120 V	10A	3 AB, 10A	75 k, 1/2W	RCA-2N3669
240 V	1.5A	3 AG, 2A, Quick Act	150 k, 1/2W	RCA-2N3529
240 V	5A	3 AB, 5A	150 k, 1/2W	RCA-2N3525
240 V	10A	3 AB, 10A	150 k, 1/2W	RCA-2N3670

4-23 Full-wave motor control circuit using antiparallel-connected SCRs. RCA

pulse of forward bias is impressed at one base, an amplified pulse of forward bias appears at the other base. In turn, the first base receives reinforced forward bias, etc. The circuit experiences a cumulative sequence of actions leading to a locked condition with both transistors heavily saturated in their conductive states. Because this process is regenerative, the transition from the off to the on state is almost instantaneous and is accompanied by a sharp pulse that is suitable for triggering SCRs. It is not even necessary to initiate the action with a pulse—a slowly rising signal level suffices. In the actual circuit of Fig. 4-23, the slowly rising (relatively speaking) base signal is obtained from the RC phase-shift network comprising R1, R2, and C1. By adjusting R2, the phase of the voltage derived from this network can be varied, thereby controlling the time of triggering of the SCRs, which, in turn, determines the effective value of the voltage applied to the motor.

This system is best suited to the control of large universal motors. Series-type dc motors can sometimes be successfully used up to 60 Hz. At higher frequencies, there would be high losses and also commutation problems. Permanent-magnet and shunt motors cannot be operated from sources providing ac waveforms such as this controller provides. A split-phase induction motor of the permanent-capacitor type can, surprisingly, undergo considerable speed variation when energized from this controller; this is particularly true when the motor load is a fan. (It is also true that fan performance changes greatly for relatively small speed changes.) The experimenter might wish to use a controller of this type with shaded-pole induction motors, although these are not often encountered in large sizes due to their inherently low starting torque.

Almost any NPN-PNP small-signal transistors will serve in the trigger circuit, and little is gained by striving for a close match in transistor parameters. However, it is often wise to use overrated components in industrial equipment where RFI, EMI, and transients tend to be much stronger than is generally encountered in communications or instrumentation circuits. Thus, a pair of transistors with higher voltage and current capability than is ordinarily identified with small transistors would be suitable for this controller; such a pair of transistors is the RCA1A18 NPN type, and the RCA1A19 PNP type. These devices have 1 A maximum current ratings and are available in the JEDEC TO-39 package. Heat sinking is not necessary because of the relatively low dissipation incurred in the regenerative switching process.

There is no performance penalty associated with this trigger circuit. The controller provides a phase adjustment range of 5 to 170 degrees; for most practical purposes, this may be said to correspond to a power control range of zero to 100 percent, where 100 percent is the power that the motor would extract directly from the utility line.

Speed control of larger induction motors

The motor-control techniques discussed pertain to motors in which speed can be controlled or stabilized by varying the voltage or current applied to them. This has included nearly every type of motor except ordinary induction motors. In some instances, very small induction motors with wound rotors, split-phase types (especially the permanent capacitor split-phase motor), and shaded-pole types can be controlled by voltage; the speed variation is relatively limited and is best realized when the load is a fan or blower. Textbooks dealing with induction motors often ignore these exceptions, simply labelling induction motors as constant-speed devices. Of course, the induction motor does shed speed with load, but the implication is that the slip or departure from synchronous speed (the rotational speed of the magnetic field) is only a few percent, and moreover is not primarily influenced by voltage or current.

The induction motor is the workhorse of industry, especially in its three-phase format. The three-phase induction motor is self-starting, very efficient, easily reversed, electrically simple and mechanically rugged, and displays desirable torque and speed characteristics for many applications. A tempered statement applies to single-phase induction motors with the major exception being that they are not inherently self-starting. Although there is no magic line of demarcation, three-phase motors are largely found in the integral-horsepower ranges. Single-phase induction motors for heavy-duty uses feature cast rotors (as do the three-phase types) and have fractional-horsepower ratings from about one tenth to one horsepower. Such ordinary induction motors are widely used but are not amenable to speed control by any of the techniques thus far described.

The induction motors that can be controlled, at least to some extent, by some of the previously described circuits are generally smaller than one tenth horsepower and are of the specialized types alluded to above. Control should only be attempted with circuitry supplying full-wave ac waveforms to the motor.

In the past, the only practical (but expensive) way to vary the speed of the three-phase or ordinary single-phase motor was via an electromechanical frequency changer—this generally comprised an alternator driven either by an internal-combustion engine, or by a dc motor. With the advent of solid-state power devices, a new approach has seen considerable development. It remains understood that the supply frequency must be controlled, but this is now done by means of electronic inverters. This has endowed industrial processes with the basic desirable features of induction motors with the added advantage that these have, in essence, become variable-speed machines. The initial investment is a fraction of the cost of the mechanical approach. Moreover, the electronic inverter and induction motor combination is admirably suited for traction vehicles, where the maintenance of brush-commutator machines can be entirely circumvented.

In technical literature dealing with the control of large induction motors, it is commonplace to use block diagrams, functional schematics, and simplified and partial circuits more extensively than is customary with applications appealing more to hobbyists and experimenters. It is assumed that control of larger motors is more the domain of professional designers and the important thing to convey is the basic principle, rather than details pertaining to parts, devices, and localized circuits. These motor controllers tend to be considerably more complex than those that have been discussed for smaller motors. Also, this voltage-current region is a hazardous one where the average electronics practitioner's work habits are not always compatible with safety of person and equipment. And, finally, this is a subject unto itself—a specialty requiring voluminous treatment not within the scope of this book. However, a brief allusion will be made to the important aspects of large induction motor control because of the necessarily nebulous distinction between large and small induction motors.

To start with, realize that the average 60 Hz induction motor is inherently capable of running at a fraction of its rated speed, and also at several times its rated speed. To be sure, special motors can be designed for such purposes that will do the job somewhat more efficiently, and possibly smoother at low speeds than garden-variety types. However, both practical experience and mathematical design show that ordinary motors have considerable capability for such service. This tends to be true even in many instances wherein the applied waveform is closer to square than sinusoidal, although departure from sine-wave excitation can degrade operating efficiency and can roughen the torque characteristic under certain speed/load conditions. The three-phase machine, especially, is quite forgiving; like a multicylinder engine, its inherent smoothness overcomes nonideal conditions.

There is one criterion that must be met with all variable-frequency control systems for induction motors. Except if one is content with a small range of speed control, the applied voltage to the motor should vary (either manually or automatically) directly as the frequency. Thus, if the frequency (and speed) is doubled, twice the rated (60 Hz) voltage should be impressed on the motor. Conversely, if the frequency is halved to 30 Hz in order to bring about half-speed operation, the motor should receive one half of its 60 Hz voltage. If this were not done, the motor would suffer severe loss of torque at the higher frequency, and would be severely overloaded electrically at the lower frequency. Note, however, that the change in voltage, as such, is not the speed-changing agency.

The block diagram of Fig. 4-24 depicts the general control scheme used with three-phase induction motors. It is basically a driven inverter using two output stages per phase. These output stages function as switches and can be sequenced in various ways for powering the motor. Usually, it is desirable to make the best use of both digital and analog operation. For power-handling efficiency, the output stages should have only two states—on and off. For good torque characteristics, it

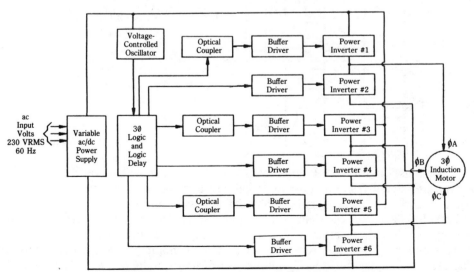

4-24 Generalized setup for speed control of large induction motors.

is best that the motor current be sinusoidal. Although wave purity is not as important as in audio amplifiers, a reasonable approximation to a sine wave results in smooth motor operation and prevents unnecessary temperature rise from eddy currents and hysteresis. Two methods of complying with the needs of the output stages and the motor have become popular. These are the six-step system and the pulse-width modulation system. Both represent formats for switching the output stages so that three-phase quasi-sine waves will be delivered to the motor. Both systems make use of motor inductance to smooth the pulsed waveforms for a better approach to a sinusoidal shape. With small variations, both methods can be functionally represented by the general setup shown in Fig. 4-24. The system, as shown, is set up for the six-step mode.

As its name implies, the six-step mode fabricates quasi-sine waves from six basic constructs—square waves with predetermined amplitudes, durations, and times of occurrence. Figure 4-25 illustrates such a three-phase format of voltages and currents. Because it is the current that develops the torque in the motor, particular attention is directed to its approximation to sinusoidal shape. The three top waveforms are phase voltages. The three lower waveforms are phase currents. In this regard, the dashed waves in Fig. 4-25 represent the motor current and show the smoothing effect of motor inductance.

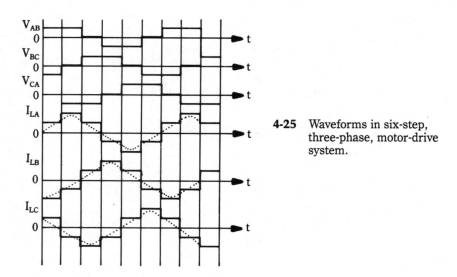

4-25   Waveforms in six-step, three-phase, motor-drive system.

The pulse-width modulation system compares a low-frequency sine wave with a high-frequency triangular wave, obtaining a wave train of varying duty-cycle square waves. This wave, again, is a quasi-sine wave insofar as concerns its circuit effects. In Fig. 4-26A the manner of bringing about this result is illustrated; either a voltage comparator, or an over-driven op amp work well in this capacity. The wave train in Fig. 4-26B represents one of the motor phase currents, and the dashed wave shows the smoothening effect of motor inductance. As with the six-step method, the best features of digital and analog technology are systematically

A

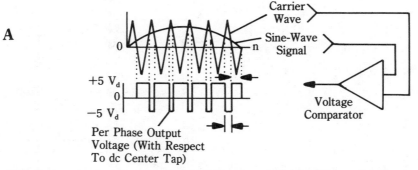

Carrier Wave

Sine-Wave Signal

Voltage Comparator

0

n

+5 V$_d$
0
−5 V$_d$

Per Phase Output
Voltage (With Respect
To dc Center Tap)

High-frequency triangular wave (carrier) compared with low-frequency sine wave produces a pulse-width modulated wave.

B

I_{LA}

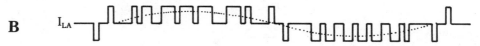

Phase-current wave in motor simulates original sine wave through agency of varying-duration pulses.

4-26   Waveforms in pulse-width modulation motor-drive system.

deployed—the output stages perform as switches, and the motor receives at least quasi-sine waves of drive currents.

More detailed circuitry of a drive system representative of the six-step wave-synthesis technique is shown in Fig. 4-27. In this case, the design is intended to control a 2 HP 230 V, three-phase induction motor. As inferred, six-step systems have much in common with pulse-width modulation systems, and the power-function blocks of the two systems tend to be readily interchangeable. This being the case, it might prove profitable to make the necessary minor changes and adaptations so that Fig. 4-27 can handle pulse-width modulated, rather than six-step, signals. It is easier to obtain smooth motor operation, especially at low speeds, with pulse-width modulation. Also, the need to vary the dc voltage supplied to the inverter can be dispensed with—although motor voltage must still be varied with speed, this can be achieved by varying the width of the pulses. And best of all, the semiconductor firms have made available ICs for attaining all of the logic required in a three-phase pulse-width modulation system with minimal complexity and expense. This has come about as the result of these companies' previous experience with pulse-width modulation ICs used in switching power supplies.

In dealing with high power levels and large induction motors, it is often desirable that the inverter be current, rather than voltage, fed. That is, the output element of the power supply should be an inductor, rather than a capacitor. This might appear to be a triviality, but it tends to enhance the reliability of the overall system. The circuit of Fig. 4-28 shows one way of accomplishing this. The SCRs in the three-phase power supply can be phase controlled in order to control motor current, and the six inverters can be driven either the six-step or PWM switching formats. This technique greatly reduces the peak currents in the output stages and

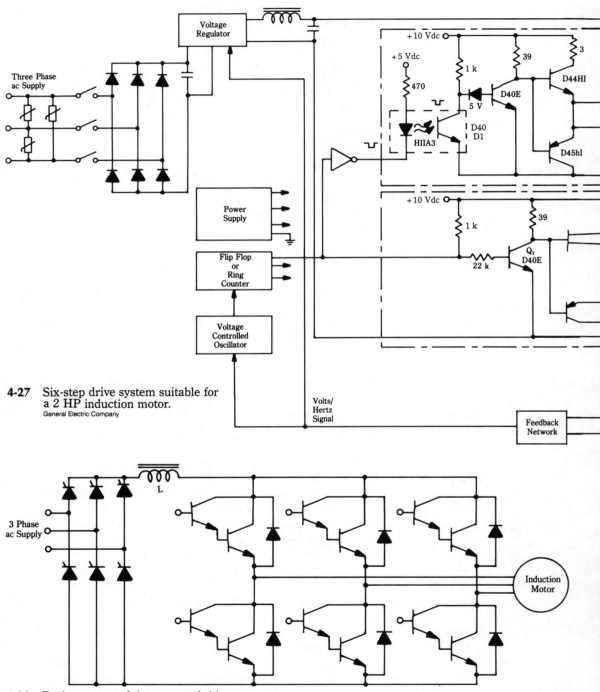

4-27 Six-step drive system suitable for a 2 HP induction motor.
General Electric Company

4-28 Basic concept of the current-fed inverter.

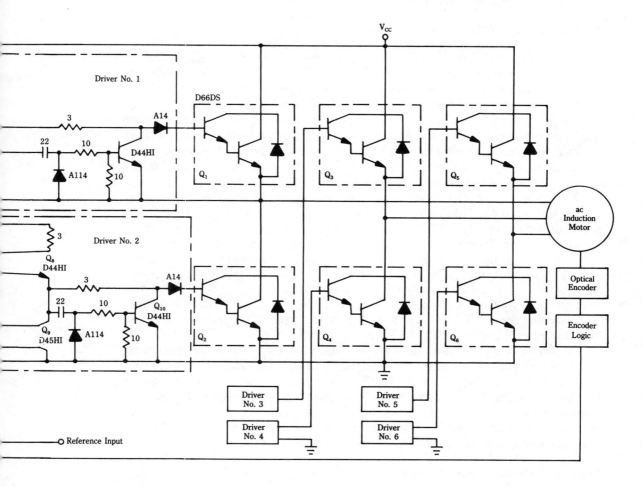

limits the tremendously high starting and blocked-rotor currents of large motors. Experience indicates that the current-driven inverter prolongs motor life. And, because switching transients are reduced in violence, RFI and EMI problems tend to be less severe than with voltage-driven inverters. A practical problem, however, is that at low frequencies, the inductor may have to be inordinately large.

Dedicated IC for controlling four-phase stepping motors

Servo systems have long been used to both control and position the shafts of motors. Various degrees of success have been attained; in principle, a high amplification of the error signal should result in arbitrarily precise position control. In practice, high gain provokes problems of overshoot, instability, and oscillation. The critically damped, mathematically obedient systems described in textbooks are not found in most applications because of unpredictable, varying, and uncontrollable inertial and frictional forces from bearings, gears, shafts, belts, etc. A better way becoming popular is through the computerized control of stepping motors.

Stepping motors advance one step at a time in response to appropriate pulses. The basic idea of control is to supply predetermined sequences of pulses. If these can be controlled in rate, time of occurrence, and format of application to the windings of the stepping motor, the direction, speed, and stopping position of the motor can be precisely put through its paces. The instabilities of servo systems, such as hunting, are minimal or absent because operation is not dependent on feedback, or amplification of an error signal. But, until recently, the electronics needed for this purpose entailed numerous discrete components in large racks and panels.

The Cybernetics Micro Systems CY525 Intelligent Ramping Stepper-Motor Controller is a 40-pin dedicated IC designed to interface a command source and a power driver for the motor. The command source can be an ASCII (American Standard Code for Information Interchange) keyboard via high-level language, or a computer or microprocessor in which instances the commands can use binary code. The salient feature of ASCII keyboard control is that a system can be prototyped prior to introducing computer commands. It is much more convenient, and usually more economical, to explore system behavior by typing in commands than by harnessing the services of a computer. Of course, many systems can be successfully implemented by retaining the keyboard control.

Performance control is almost limitless and involves primarily programming skills, rather than hardware circuitry design. Up to 10,000 steps per second can be handled. Best of all, sequences of high-level commands can be stored internally and can be executed upon command. A continuous-run mode simulates conventional motors by supplying an unlimited number of steps. When operating in this fashion, the stepping motor behaves similarly to a synchronous motor in that the rotational speed is locked to the frequency or pulse rate of its drive source.

Figure 4-29 depicts a simple stepping-motor control setup using the CY525. The ASCII high-level commands and decimal values from the keyboard are translated by the CY525 into a logic format, which is then power boosted to control the operation of the stepping motor. Power boost is accomplished by the ULN 2068B, a power Darlington transistor array made by Sprague. Other power-transistor

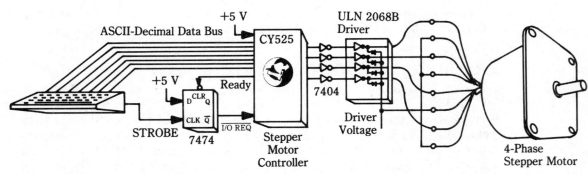

4-29   ASCII keyboard system for controlling a stepping motor using the CY525.   Cybernetics Micro Systems, Inc.

schemes can be used—the basic program being to process the logic output of the CY525 into stepped levels with sufficient power to drive the motor. In most cases, the 7404 hex inverter must be used to interface the CY525 with the power stages.

Because so many functions have been incorporated on the CY525 IC, it is only natural to wonder why the power-drive circuitry wasn't also included. Not only do nasty thermal problems assert themselves when brains and brawn are integrated on a single chip, but a point is generally reached in systems where flexibility and versatility are sacrificed if the whole system resides in a single IC. It is often better to have the system divided into several subsystems. The parts count and complexity is still dramatically reduced from what it would be with discrete-device design, but the user is accorded more options for adaptation to the unique needs of his situation.

The internal clock of the CY525 can be made operational by simply connecting a 2 to 11 MHz series-resonant crystal to pins 2 and 3. A 3.58 MHz TV colorburst crystal suffices for this purpose. However, the higher the clock frequency, the higher the maximum stepping rate. An 11 MHz crystal will allow nearly 10,000 steps per second.

The pin and logic diagrams of the CY525 are shown in Fig. 4-30. It should be obvious that the 40-pin IC comprises a very involved system, one that would dwarf the stepping motor itself if construction were attempted with discrete devices. The logic diagram also depicts in simplified form the internal architecture of the CY525. Note the PROG INSTR STORE; this enables high-level commands to be typed into memory so that they may be executed at a later time. Sixty bytes of program commands can be stored. Another salient feature of the CY525 is represented by pin 36, appropriately labelled ASCII/BIN. The logic level applied to this pin determines whether the CY525 is receptive to ASCII data or to binary arithmetic commands. Thus, a logic high enables operation from ASCII commands; a logic low makes the system responsive to binary code, such as might be received from a computer or a microprocessor.

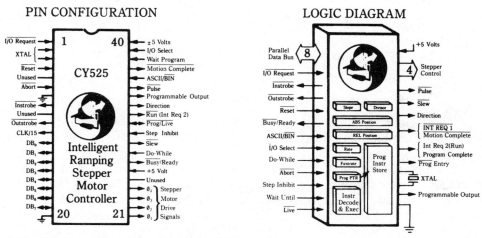

4-30 Pinout and logic diagram for the CY525 stepping motor controller. Cybernetics Micro Systems, Inc.

Table 4-1 is a list of the programming features of the CY525. A study of these capabilities should drive home the fact that versatility of control greatly exceeds what can be readily accomplished with analog servo systems. It is true that the stepping motor cannot compete with conventional motors in the brute-force parameters of torque and horsepower; within their capabilities, however, stepping motors together with logic-circuit drive systems provide superior control accuracy and repeatability. Relatively large and powerful stepping motors have been designed for special applications, and the commercial availability of larger stepping motors is now likely in view of the high-current solid-state power devices now available for pulsing these motors. Although the logic instructions needed to implement the performance features listed in Table 4-1 will not be dealt with here, putting the stepping motor through its paces should prove an intriguing project for the enterprising programmer.

Table 4-1. Programming features of the CY525 stepping-motor controller.

Programmable via ASCII keyboard	Two interrupt request outputs
ASCII-decimal or binary communication	
	Programmable output line
Single 5-volt power supply	Programmable delay
27 hi-level language commands	Verify register/buffer contents
Stored program capability (60 bytes)	
	Several sync inputs and outputs
Linear acceleration, definable	Abort capability
Change rates while stepping	Step inhibit operation
Read position on-the-fly	Ability to turn off phases
10,000 steps per second (11 MHz xtal)	
	Slewing indication output
Absolute/relative position modes	"Dowhile" and "Wait Until" commands
Define starting rate and slew rate	"Jump to" command
Ramp-up/slew/ramp-down	Loop command with repetition count
Hardware or software start/stop	Allows address labels for loop and jump commands
Software direction control	Unlimited number of steps in continuous mode

Cybernetics Micro Systems, Inc.

The command summary of the CY525 is shown in Table 4-2. Note that there are 27 of these high-level instructions for bringing about the various operating modes.

A setup for placing the CY525 in operation is shown in Fig. 4-31. This is essentially a more detailed version of Fig. 4-29. Note that LEDs are provided for visually indicating states and state transitions. The basic idea is to establish a prototype system with the aid of the ASCII keyboard. If desired thereafter, the commands can originate in an 8-bit data bus, a microcomputer, or in data stored in PROM (programmable read-only memory), EPROM (electrically programmable

**Table 4-2. Command summary
of the CY525 stepping-motor controller.**

Aa	Absolute location specified	Rr	Rate of stepping
B	Bitset (control output = 1)	Ss	Slope of accel/decel
C	Clearbit (control output = 0)	Tt	Til pin 28 HI, repeat program
Dd	Delay specified milliseconds	U	Until WAIT LO, wait here
E	Enter program into CY525	Vv	Verify buffer contents
Ff	First step rate	W	Wait until WAIT line HI
G	Go (begin stepping)	X	Execute stored program
H	Haltmode or continuous step	Zz	Divisor for slope
I	Initialize-software reset	+	Set clockwise direction
Jj	Jump to prog buffer location	–	Counterclockwise direction
Lc,a	Loop through prog segment	0	Return to command mode
Nn	Number of steps	$	Marker to Jump & Loop to
Oo	Offset drive signals as req'd		
Pp	Position to step to		These commands can be either stored or executed.
Q	Quit entering program code		

Cybernetics Micro Systems, Inc.

read-only memory), or ROM (read-only memory). This prototyping capability greatly simplifies the tailoring of stepping-motor control to the requirements and limitations of a practical system.

Electronically upgrading the efficiency of induction motors

Engineers exposed to electrical engineering curricula prior to the availability of solid-state power devices and control ICs had it easy; the electric motors studied were seemingly endowed with built-in behaviors. Textbooks devoted pages to rigorous explanations of causes and effects, documented by profuse mathematical support. Now is a different era. It is true that the basic motors are still very much here, and it remains true that the natural characteristics of a particular motor can optimize its use for a certain application. To a large extent, however, it is now feasible to considerably modify the performance of a motor via electronic control. This means that there is now much overlap among motor types—operating parameters such as speed, torque, regulation, starting capability, etc., can be manipulated at will.

An interesting example of such electronic manipulation of basic characteristics is a useful technique that can be readily applied to the single-phase fractional-horsepower induction motor. These motors are used by the many millions, not only in industry but for household appliances. A bare-bones induction motor is not self-starting from a single-phase line. However, the preponderance of commercial models are self-starting by incorporation of a centrifugal-switching scheme that momentarily causes the motor to behave as a two-phase machine. Various ways of doing this have been devised; most make use of an additional stator winding, located 90 electrical degrees from the main, or *running* stator winding. Either

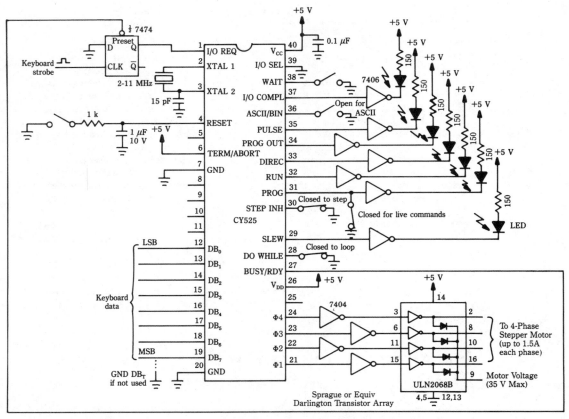

4-31   Setup for placing the CY525 stepping-motor controller in operation. Cybernetics Micro Systems, Inc.

resistance or capacitance can be used in conjunction with the starting winding to provide sufficient phase shift to start the motor as a two-phase machine. As the motor approaches its operating speed, the centrifugal switch disconnects the starting winding.

Except that it is a rotational device, such a motor behaves as a static transformer. In particular, it attains its highest efficiency at or near rated load. The load, of course, is a mechanical one. At light loads, its efficiency drops off considerably; that is, it consumes more power from the utility line than would otherwise be the case. The reason for this is that at heavy loads, the motor looks resistive to the ac line. At light loads, the motor becomes inductive to the extent that the power factor is appreciably lowered. Low power factor always implies greater current consumption from the ac line than would prevail at, or near unity power factor. For most of the long history of this workhorse motor, such behavior was accepted as the nature of the beast.

The cost implication of this behavior is obvious enough. But there also are other disadvantages; the motor will necessarily run hotter and often noisier. There is, indirectly, greater stress on the bearings.

If the applied ac voltage to a lightly loaded induction motor is manually lowered, say by means of a Variac, it can be observed that the motor will then begin to appear more resistive and less inductive. That is, its power factor will improve (become higher) and its ac line current will decrease. Of course, the applied line voltage cannot be reduced too much, or the motor will stall. Such an observation has great significance. It tells you that the lightly loaded motor has an excessive appetite for current. It will remain functional and will be in better health at reduced current. The objective must be to bring down the allowable motor current when the motor becomes lightly loaded.

From the foregoing, it should be apparent that the solution to this problem begins with the sensing of the phase displacement between applied voltage and the resultant motor current. This is true because power factor is a manifestation of phase. At low loading, the power factor of the motor becomes low as consequence of the increased phase displacement between the applied voltage and the resultant motor current. Once this phase displacement is sensed, it becomes rather straightforward to cause a triac to lower current consumption by decreasing the impressed voltage; in turn, this forces the motor to operate at an improved power factor.

Many circuit techniques have been devised to accomplish this automatic control. However, it is advisable to the experimenter to get one of the dedicated ICs now available from various vendors. These ICs require minimal additional parts, in addition to the external triac. Not only does this procedure avoid inevitable headaches and catastrophes, but it enables sufficient compactness to be installed within the motor bell housing in many instances. In this way, you circumvent nasty problems having to do with insulation, current and power ratings, and instability (it is, after all, a feedback system). In cases like this, it is plainly pragmatic to capitalize on the experiences and creativity of engineers who have devoted many hours to perfect a difficult function. Aside from the basic techniques of sensing phase and controlling the firing time of a triac, there are many subtleties involving starting, stability, transient loads, and harmonic energy that cannot readily be coped with by the experimenter working with op amps, logic circuits, and discrete devices.

The block diagram of the basic energy saver for induction motors is shown in Fig. 4-32. The circuitry comprised of the three logic gates can be said to be the heart of the control system. Specifically, the output of the OR gate is a pulse train having widths proportional to the phase angle between applied motor voltage and the resultant motor current. Motor current is sensed across a low resistance in series with the motor; this resistance is low enough to cause negligible power dissipation or effect on the operation of the motor. Two current-squaring amplifiers, as well as two voltage-squaring amplifiers are needed to satisfy the input requirements of the AND gates.

The summing amplifier provides a means of adjusting the degree of automatic control that will be optimum for the particular motor and its application. Overcontrol can lead to instability of motor operation. The smoothened error-voltage from the summing amplifier is not yet, however, ready for application to the gate of the triac because it is necessary to synchronize the control with the ac line voltage. This matter is taken care of by a line-synchronized ramp generator and a comparator. The output of the comparator is again a wave train of pulses. This time,

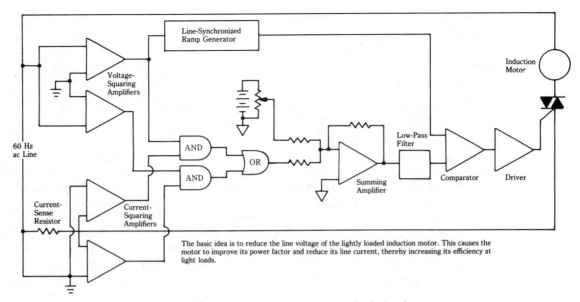

The basic idea is to reduce the line voltage of the lightly loaded induction motor. This causes the motor to improve its power factor and reduce its line current, thereby increasing its efficiency at light loads.

4-32   Block diagram of energy saver for induction motors.

however, the pulses are position modulated so that the triac triggering is synchronized to the line voltage. You can appreciate that the pulsed wave train directly from the summing amplifier would produce random triggering of the triac, and could not bring about the intended function of the control system.

Figure 4-33 shows the basic implementation of a dedicated IC, the Harris HV-1000/1000A for enhancing the efficiency of an induction motor. Known as an induction motor energy saver, this IC obviously simplifies the above described control technique. The triac should be sized to safely accommodate about three times normal full-load current for the motor. This safeguards against damage from locked-rotor current, which is encountered at times in many applications. This dedicated IC incorporates refinements not depicted in the simplified block diagram, such as anticipating loads, the ability to handle shock loads, and provision for optimum starting conditions.

Clean control technique for three-phase motors

The mere turning on and off a three-phase fractional or integral-horsepower motor is not a trivial matter. And switching accommodation for reversing the rotation further aggravates problems when conventional mechanical switches are used to make and break line currents. Because of the large current-handling capacity needed, mechanical or electro-mechanical switches are bulky and costly. Moreover, maintenance is necessary to take care of deterioration resulting from arcing, sparking, heating, corrosion, and wear. A particularly bad feature of switching

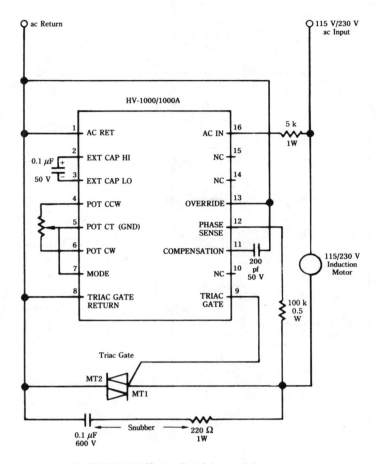

The HV-1000/1000A IC makes this technique practical.

4-33   Basic implementation of the induction motor energy saver. Harris Semiconductor Corp.

motor currents with physical contacts is that the arcing and sparking generates severe EMI and RFI. Such interference not only incapacitates radio, TV, and communications systems, but can produce inadvertent triggering of industrial control circuits, as well.

The scheme shown in Fig. 4-34 enables control of a three-phase motor from signal- or logic-level voltages and currents. The actual motor currents are turned on or off electronically by power triacs so that arcing or sparking does not occur. Moreover, the actual switching intervals take place at times of zero line voltage, further reducing the possibility of line transients. This switching mode prolongs motor life by sparing it from heavy current inrushes. Another nice feature of this arrangement is that the rotation of the motor can be readily reversed. Advantage is taken of the fact that the transposition of any two leads to a three-phase motor reverses its rotation.

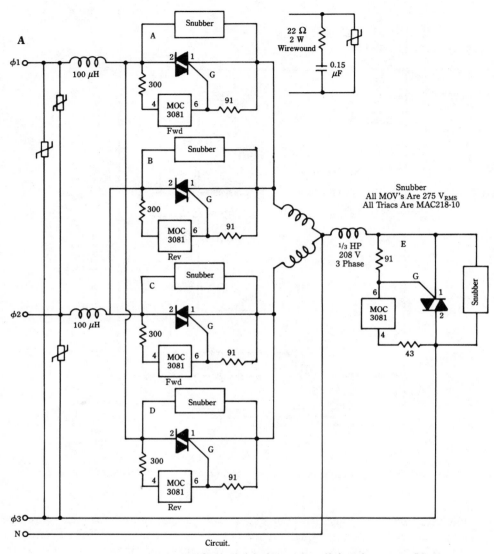

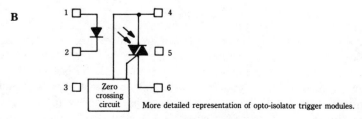

4-34 Electronic control scheme for a three-phase motor. Motorola Semiconductor Products, Inc.

When the motor is on, three of the five power triacs will be active. Triac E will be active whether motor rotation is forward or reverse. For forward rotation, power triacs A and C will also be active. For reverse rotation, however, these two power triacs will be inactive, but power triacs B and D will then be active. These commands are executed by providing input currents to the appropriate optoisolators (terminals 1 and 2 of the MOC 3081s). The motor is turned off by depriving all of the optoisolators of input current. Either a logic system or small switches or relays may be used to control the currents applied to the inputs of the optoisolators. An idiot-proof method should be implemented to prevent the wrong pattern of power triacs from being activated.

Note that the snubber networks include MOV transient-energy absorbers. Thus, there is a total of eight MOVs, all of the same rating, used in the overall circuit.

Building-block drive scheme for electric vehicles

Direct-current motors have desirable characteristics for electric vehicles. Next to basic considerations of speed control, power, and torque, the all-important need is for high efficiency in order to extend battery capability between chargings. One way of improving efficiency is to use pulse-width modulation or duty-cycle variation in place of simple rheostatic (variable resistance) control of the armature current. Via such switching techniques, a great deal of I^2R dissipative loss can be eliminated.

An additional expedient for conserving battery drain is to use regenerative braking. The basic idea here is to use the drive motor as a generator during slow down. Not only is energy returned to the battery during such intervals, but the natural counter torque of the motor-turned-generator assists actual braking of the vehicle.

It happens that complementary-symmetry power MOSFETs can be arranged in a simple configuration to meet the needs of a PWM system with regenerative braking. Such power MOSFETs are easy to drive, easy to parallel, electrically rugged, and can have both minimal switching and conductive losses. Moreover, their intrinsic diodes can perform useful functions in the circuit, thereby reducing both parts count and cost. Viewed as a basic building block of an electric vehicle drive system, the representative circuit is shown in Fig. 4-35. A nice feature of this arrangement is that there is no possibility of simultaneous conduction in the power MOSFETs—both the N-channel unit and the P-channel unit cease conduction when the alternate unit is turned on.

The operation is as follows. In the motoring mode, the N-channel MOSFET Q1 is chopped at a suitable pulse-repetition rate and duty cycle to give the vehicle desired motion. As long as acceleration or constant speed is desired, the P-channel power MOSFET Q2 is kept in its nonconducting state. However, the intrinsic diode of Q2 serves as the free-wheeling diode to maintain steady current flow through the motor armature despite the chopping action of Q1.

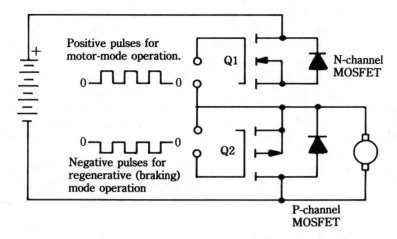

Positive pulses for motor-mode operation.

0 ⊓⊔⊓⊔⊓⊔ 0

Q1

N-channel MOSFET

0 ⊔⊓⊔⊓⊔⊓ 0

Q2

Negative pulses for regenerative (braking) mode operation

P-channel MOSFET

Each of the complementary-symmetry MOSFETS can be paralleled with additional units in order to satisfy current requirements.

4-35   Circuit of building block for driving electric vehicles.

Conversely, when it is desired to decelerate the vehicle, Q2 is chopped at a suitable rate, and Q1 is kept in its nonconducting state. The intrinsic diode of Q1 then allows current from the generator to complete its path through the battery, thereby charging it. As alluded to before, this return of energy is accompanied by electromagnetic braking action in the armature. As can be seen, two accomplishments are realized through this circuit action—enhanced braking is obtained for the vehicle, and charging current that would be otherwise wasted is returned to the battery.

Despite the topographic simplicity of this motor-driven circuit, its behavior is not readily obvious from inspection. Indeed, it might seem that the two free-wheeling diodes are not properly polarized to participate in the circuit sequences already described. In order to see why the operation is as outlined, it is necessary to keep in mind that the power MOSFETs are never in a sustained ON state. Rather, each power MOSFET is chopped. Also, when one power MOSFET is chopped, the other is OFF or inactive. It is the chopping action that gives the circuit a different mode of operation than would be the case if either power MOSFET was permanently turned on.

In order to obtain clearer insight into circuit operation, assume the motor to be replaced by an approximately-equivalent network comprising an inductance, a resistance, and a capacitance, as shown in Fig. 4-36. The basis for this network is that the inductance of the motor armature can be simulated by a series inductance. The rest of the armature then appears as a load resistance for motor action or as the internal resistance of a voltage source for generator action. The capacitor maintains voltage across the load and roughly simulates the generator action of the motor, which is present whether the motor is receiving energy (motor mode) or is delivering energy (generator mode).

The helpful aspect of this equivalent motor circuit is that it duplicates the out-

A

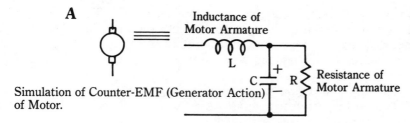

Simulation of Counter-EMF (Generator Action) of Motor.

Approximate equivalent circuit of a dc motor.

B

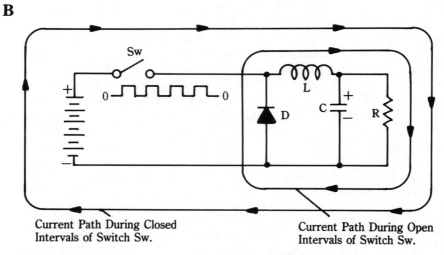

Current Path During Closed Intervals of Switch Sw.

Current Path During Open Intervals of Switch Sw.

With system in motoring mode, the circuit resembles and behaves similarly to a buck-switching power supply.

C

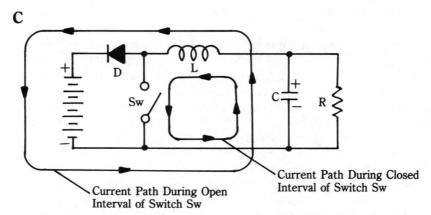

Current Path During Open Interval of Switch Sw

Current Path During Closed Interval of Switch Sw

With system in regenerative, or braking, mode, current flows through diode and charges battery. In this case, assume C is very large and has been previously charged to near battery voltage. Also, assume R is very high.

4-36 Aids in understanding of operation of vehicle motor circuit.

put circuit of switch-mode power supplies. The greater your familiarity with the operation of most switch-mode supplies—those with an output circuit comprising a series inductance and a load made up of a resistance and an output filter capacitance in parallel—the easier it will be to understand why the motor drive circuit operates as it does.

The key to such understanding is to realize that current through the inductance tries to continue flowing after the switches open. In order to do so, the current paths then find completion through the respective free-wheeling diodes. In this way, the excess energy stored in the inductance depletes itself by sustaining continuity of motor torque (in the motor mode of operation) or by charging the battery (in the generator or braking mode of operation).

Referring to Fig. 4-36A, the dc motor is depicted as an approximately equivalent network having the same topography as buck-type switching regulator output circuits. The capacitor here simulates the counter-EMF (generator action) of a motor. For the purpose at hand, assume the capacitor to be very large and the resistance very high. Once charged, such a capacitor can supply current to the inductor; that is, it, like the generated voltage from a free-spinning or mechanically driven motor, can act as a voltage source. This analogy will be found useful in understanding the behavior of the system when it is in its regenerative mode of operation.

When the system is in its motor mode of operation, the situation is closely simulated by the circuit shown in B of Fig. 4-36. This is essentially the circuit action that occurs in the output circuit of a buck-type switching regulator. When the switch is in its OFF interval, the current is maintained through the inductor by taking the path through the free-wheeling diode. This current is now produced by the collapsing magnetic field of the inductor. In other words, energy stored in the magnetic field of the inductor during ON time, is now transferred to the electric field, giving rise to an EMF that is so polarized as to force current through the free-wheeling diode. The net result is a near-steady, rather than a pulsating, current through the load.

The gist of the above explanation is that this is the way circuit behavior occurs in the actual motor circuit and is primarily because of the equivalent series inductance of the motor armature. The capacitor has a relatively minor roll other than imparting even greater smoothing action to the load current.

The situation prevailing for the regenerative mode of operation is represented by the network shown in C of Fig. 4-36. In this case, it is profitable to focus on the capacitor. Assume it is very large and can supply considerable energy. Thus, it becomes a voltage source that can sustain current flow in the conductor when the switch opens. In so doing, it also establishes a new current path through the battery, thereby charging it. Because of *flyback* action at the opening of the switch the available voltage is higher than the battery voltage, a necessity for charging.

Note that the system works because of the chopping action. Current can be controlled in both modes of operation by varying the duty cycle of the chopping wave. Also, keep in mind that when one chopping sequence is under way, the other is inactive—the system operates in either the motor or the regenerative mode, never in both simultaneously.

The generalized symbol for dc motors has been used in the diagrams. The intended inference is that the system is usable with all of the common dc motor types—permanent magnet, series, and shunt. The series motor has long been a favorite for electric vehicles because of its high starting torque. However, modern permanent-magnet motors can also be respectable performers at start up. Because the series motor has obvious physical inductance in series with the armature, its operation in the motor-drive system is easier to grasp—an equivalent circuit might not be needed. Because the inductance of the series winding is relatively high (compared to armature inductance), a lower switching rate may suffice for control purposes. The reversible feature of some series motors is also a plus for vehicle application.

Although ordinary power MOSFETs are shown in Fig. 4-35, the newer IGBT (insulated-gate bipolar transistor) types merit consideration. These devices feature the input characteristics of power MOSFETs, and approximate the high output conductivity of bipolar transistors. Their combined losses (switching plus conductivity) in a motor-control system tends to be considerably less than can be attained with either power MOSFETs or bipolar power transistors. The higher operating efficiency translates into less battery drain, so that more miles between chargings can be realized. These devices are available in both N-channel and P-channel types, although less choice is available in the latter.

Power interface between logic and action

Microprocessors and other digital logic can be likened to the brains and nervous systems of biological species. There must also be an analogy to the muscles and limbs that exert specified patterns of work on command. Such an analogy can be found in the form of power semiconductors and the motors, solenoids, and other devices they actuate. A widely used power semiconductor for such purposes is the triac. This is a very useful device because powerful ac motors and solenoids can be turned on and off by commands originating in the programmed behavior of the digital circuit. The problem in such an association is the interface between logic and power circuitry.

An obvious approach is the brute-force one utilizing whatever combination of passive and active components is needed to translate logic levels to triac gate signals. A typical circuit arrangement for accomplishing this is shown in Fig. 4-37. Such methods are straightforward and have often been entirely satisfactory. A shortcoming, however, is that there is no electrical isolation between low- and high-power sections of the system. Isolation is desirable, for it helps attain operator safety and it tends to protect the logic circuitry from damaging transients. It would also block a path where transients can back up into the logic gates and cause faulty operation. Power circuits are often the victims of catastrophic failures; a common one is the burn out of lamp filaments. During such an event, the metallic vapor provides a near short-circuit current path. If this should blow out the triac, the resultant current surges could, in some instances, also damage the logic circuitry. The possibility of overload and failure in a motor-control system is even more obvious.

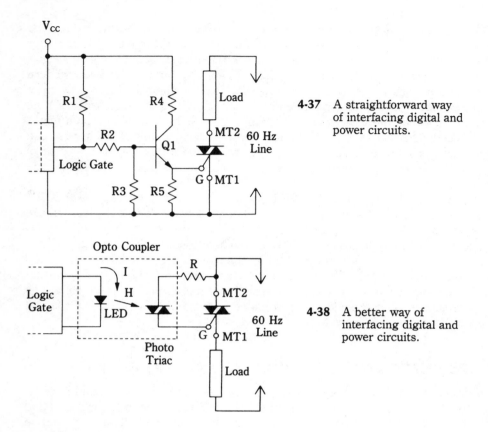

4-37 A straightforward way of interfacing digital and power circuits.

4-38 A better way of interfacing digital and power circuits.

All things considered, a nonelectrical link between logic and power systems is called for.

Figure 4-38 depicts a simplified scheme utilizing an optical link between logic and power. Here, an LED actuates a phototriac which, in turn, triggers the main triac. Basically, the phototriac substitutes for the commonly used diac in triac trigger applications. It is obvious that the alluded-to unilateral transmission path is complete realized—no electrical event in the power circuit can be communicated back to the logic system. Although other types of optical isolators can be used, the phototriac is particularly well adapted for triggering the fairly large triacs that are often associated with the load.

5
A variety of useful applications

FOLLOWING ARE SOME APPLICATIONS OF POWER CONTROL THAT SHOULD especially appeal to the experimentally inclined and to the natural innovator. These are not blue-sky projects. They have been breadboarded, investigated, and proven technically feasible. For a number of reasons, however, some have not yet attained widespread recognition as everyday techniques for controlling power. Some will probably become more popular as problems of packaging, pricing, production yield, and publicizing are resolved. In other instances, the promise is obviously there, but there remain technical drawbacks to overcome. Also, a strong factor in slowing acceptance of the newer techniques is best identified as psychological—there seems to be a natural reluctance to discard prevailing practices in favor of some new way of doing things. This holds true in spite of the merits of newer methods.

It is admittedly true that some of the off-beat techniques have clearly identifiable bugs in them. But it is also true that an evolutionary process tends to overcome and eliminate defective operation once production and marketing get started. Stereo FM broadcasting was such an example. So was varactor turning of TV sets. So were switching power supplies. Further delaying acceptance of new techniques and devices are the misleading and mistaken predictions of experts in the field who seem to miss the boat in much the same manner that persons well-versed in finance and economics misread stock-market trends. Those who have accumulated some years and retain long memories can recall the predictions of otherwise-reliable technology prophets who didn't foresee much chance for the transistor ever competing with tubes in electronics circuits—well, maybe for toys and low-level applications, but certainly not where respectable power levels must be controlled!

Mindful of these, and numerous other thumbs-down evaluations of those who should know better, some of the following circuits might, indeed, be diamonds in the rough. The intuitive experimenter, as well as the analytical designer, might be able to incorporate some modification or refinement that will make it compelling for all concerned to welcome new approaches to power control.

Because of their essentially experimental nature, some of these circuits will be presented in simplified form, generally without component values and without much implemental guidance or how-to-build discussions. The basic objective in these instances is to present the unique principles underlying the operation of power-control techniques that have not yet become garden variety.

Other circuits presented are time tested and sure fire. They are a potpourri of useful power-control applications that did not qualify for inclusion under the titles or dominant themes covered in previous chapters.

Remote switching of ac-powered loads

A commonly encountered control situation involves a lamp, motor, or appliance and a nearby source of ac utility power, but a remotely located point from which it is desired to execute switching control. Such a circumstance is to be found in both industrial and residential environments. The traditional solution via a run of conductors in conduit is often both expensive and inconvenient. In order to properly comply with building codes and insurance stipulations, such work is generally done by an electrical contractor. Together with skilled labor and often formidable wall, ceiling, or floor surgery, such as installation may prove to be far in excess of the apparent triviality of adding a couple of conductors for extending a switching function. Two ways of possibly circumventing a formal conduit-type installation are shown in Figs. 5-1 and 5-2.

In Fig. 5-1, the interface device is the Motorola MOC3011, an optically coupled triac driver. It contains an LED infrared emitter and a gateless triac device which is optically actuated. This internal triac then controls the gate of the large external triac associated with the load. The MOC3011 can withstand 7500 V, so it should be well suited to an application of this nature. Nonetheless, it would be good practice to electrically ground one of the #22 wires. This must be done in such a

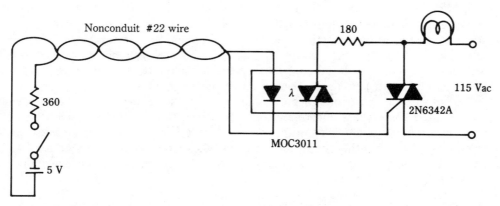

5-1   Remote control of distant ac load by isolated run of small wire. Motorola Semiconductor Products, Inc.

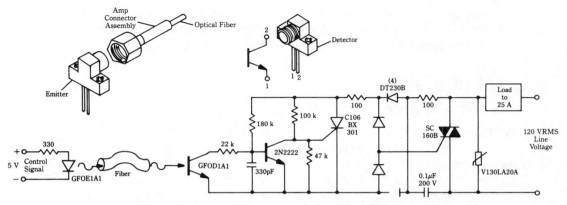

5-2 Remote control of distant ac load via optical fiber. General Electric Co.

way that there is no ground conflict with the 5 V dc source. (This probably will not be a battery as depicted.) The optocoupler should not be exposed to the weather. The MOC3011 can supply as much as 100 mA for the main triac gate. The 2N6342A triac shown in Fig. 5-1 is a 12 A, 200 V device.

A different implementation of optical coupling for the same problem is shown in Fig. 5-2. Here, the physical separation between the ac-powered load and the control point is made with a run of optical fiber. Because of the optical hardware built into the General Electric optical emitter and optical detector, the installation of this system is much easier than it is when it is necessary to obtain compatible connector elements which, at the same time, must be optically efficient.

Note that the 2N2222 NPN transistor is normally in its conducting state, thereby depriving the SCR of gate current. When light from the fiber impinges on the phototransistor detector, the 2N2222 transistor is turned off; in turn, the SCR receives sufficient gate current to become active. This causes the triac to be turned on and ac is applied to the load. This might not be obvious from the topography of the schematic. However, the four diodes can be redrawn as a recognizable bridge, in which case it will be seen that activation of the SCR also causes current to flow in the gate circuit of the triac. Load and line transients are absorbed by the metal-oxide varistor connected across the triac.

The allowable length of the optical fiber run depends greatly upon the optical attenuation in the fiber. In evaluating manufacturers' specifications on attenuation in optical fiber, make certain that it applies to the 940-nanometer wavelength involved in this application. The experimenter might also wish to investigate the possibility of using a fiber bundle, but must then be prepared to cope with the special connectors or adapters required. Optical fibers operate in various transmission modes and are made of different materials. Efforts to standardize fibers and connectors have only partially succeeded. It is suggested that the circuit of Fig. 5-2 be made operative with a short length of a selected fiber; then experimental data can be obtained for greater lengths. In this way, failure of control can be attributed to optical-fiber attenuation, and not to the electronics of the circuit. It is often discovered that apparently excessive attenuation in the fiber actually stems from alignment problems at the connectors. For this reason, it is unwise to use any splicing techniques when the system is first being checked out for performance.

Integrated Hall-effect device for switching power

The Hall effect is simple enough. Since the late nineteenth century, it has been known that a voltage can be detected across the side surfaces of a current-carrying conductor if a magnetic field is impressed perpendicular to the path of current flow. This Hall voltage is proportional to current if the magnetic field is held constant. Similarly, the Hall voltage is proportional to magnetic-field strength of the current if held constant. Obviously, a cause-and-effect relationship exists which could have some interesting and possibly useful applications. However, an inhibiting factor to widespread commercial exploitation was the relatively minute voltage exhibited by ordinary metals. Noise, temperature effects, and the need for costly and often unreliable supporting circuitry were generally cited as valid reasons for avoiding devices based upon the Hall effect.

This practice has now been dramatically changed, for solid-state Hall elements, together with modern amplifier and logic devices, have brought high performance and reliable Hall-effect devices to the consumer market. A recent advance in this technology has succeeded in integrating the Hall element and the solid-state electronics on a single chip. It is easy to provide a simple interface circuit so that a large triac can be turned on or off. Practical implementation simply involves a physical situation where a small magnet approaches the IC to produce the desired switching function, similar to applications using a reed relay. However, the Hall device experiences no mechanical wear, and can switch at a rate of 100,000 times per second. Obviously, the switching performance is free of the contact bounce that plagues electromagnetic or other mechanical switches.

Typical of the new breed of monolithically integrated Hall-effect devices is the Sprague UGS-3020 Hall-effect digital switch. The block diagram of Fig. 5-3 reveals that in addition to the Hall-effect element X, this device contains a voltage regulator, an amplifier, a Schmitt-trigger circuit, and two independent output stages. The Schmitt trigger gives the switching function a precise amount of hysteresis so that there is a definite on and off action, thereby preventing oscillation or instability. Hysteresis also gives the switch high immunity to electrical noise and transients. A simple setup for demonstrating the switching action is shown in Fig. 5-4. Magnet speed is not a factor.

This device is intended for service in the real world. It will operate from any dc source over a range of 4.5 to 24 V, and its output stage will sink 25 mA. The constant-amplitude output is compatible with all logic families. A small permanent magnet suffices for the actuating element. The switching points are reasonably constant over a wide temperature range; operation can be had from −40 to +125°C. Strong magnetic fields are not damaging.

A control scheme for switching ac power by varying the proximity of a small

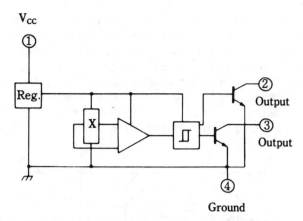

5-3 Functional block diagram of the UGS-3020 Hall-effect switch. <small>Sprague Electric Company</small>

5-4 Demonstration of switching behavior in a Hall-effect sensor. <small>Sprague Electric Company</small>

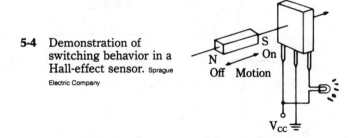

permanent magnet to the Hall-effect sensor is shown in Fig. 5-5. Here, a garden-variety optoisolator interfaces the UGS-3020 and the triac. Only one output of the UGS-3020 is used. The triac, an RCA-40669, has 8 A continuous-current capability when working from a 60 Hz, 120 V power line. The rudimentary dc power supply utilizing a 6.3 V filament transformer is more than adequate for this system; it can, in fact, be used to supply operating dc to several additional similar systems.

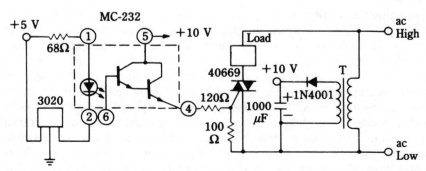

5-5 Using a Hall-effect sensor to switch an electrically isolated load. <small>Sprague Electric Company</small>

Practical insight into the sensitivity of the Hall-effect sensor can be gained from an inspection of Fig. 5-6. The south pole of the small Alnico-V rod magnet must face the brand insignia on the face of the Hall-effect sensor. Other than the indicated translatory motion can be used. For example, sliding or rotational movements sometimes better fit the mechanical aspects of the system. In any event, the speed at which the magnet moves does not affect switching performance. There is a general tendency for reduction of sensitivity with increasing temperature. This, however, is not pronounced, and in most installations will not need any special attention.

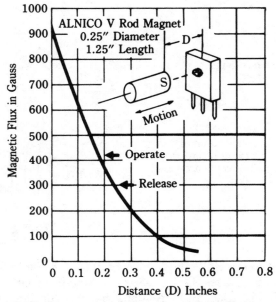

5-6 Graph depicting the sensitivity of the Hall-effect sensor.

A ruggedized solid-state relay

For experimental work and commonly encountered laboratory use, a solid-state relay should display the electrical ruggedness usually associated with electromagnetic types. However, the optoisolators generally used in solid-state relays are quite vulnerable to damage. Reverse voltage or excessive forward voltage can easily be catastrophic to the gallium-arsenide LED in these units. Overvoltage, and the resultant overcurrent can deplete the performance of LEDs even if they are not immediately destroyed. All LEDs lose light emissivity during their admittedly long useful life; moreover, useful life can be greatly shortened via abusive operating current. Accordingly, the input circuit of a solid-state relay that is not permanently designed into a circuit or system should have protective circuitry for its input.

Solid-state relays often comprise an optoisolator working into a traic. The triac is subject to increased stress when interrupting inductive loads such as motors, magnets, etc. What happens is that because of the phase displacement between current and voltage, the triac fails to commutate properly at zero-current crossings because the voltage is not also zero. This situation will be tolerated by the triac if the rate of change of voltage is slowed down. This is the function of so-called snubber networks. Ideally, a snubber network should be tailored to a particular inductive load. Practically, however, good snubber action can be attained for a reasonably wide variety of inductive loads.

Figure 5-7 shows a ruggedized solid-state relay with both input and output protection. Protection against reversed polarity of input voltage is provided by the 1N4002 diode. Protection against excessive forward-polarized input voltage is provided by the transistor connected in the circuit as a shunt voltage regulator; the higher the input voltage, the higher is the LED current, which in turn develops higher forward bias for the transistor, thereby increasing its shunting action. The transistor, by controlling the voltage drop across the 150-ohm series resistor, limits the applied voltage to the LED within safe limits.

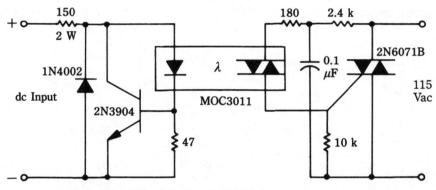

5-7 Ruggedized solid-state relay. Motorola Semiconductor Products, Inc.

The output circuit of this solid-state relay has a snubber network consisting of the 2.4 k ohm resistor and the 0.1 μF capacitor. This snubber network protects both the main triac and the drive triac from the effects of inductive loads. It also helps prevent inadvertent triggering of the main triac by power-line transients.

Solid-state relays for switching dc loads

Solid-state relays for switching dc circuits advantageously use the low saturation voltage drop of bipolar transistors. This drop is much lower than the forward voltage drop across a PN diode carrying the same current. If you were not aware of this, it would seem that the facts should be just otherwise because two PN junctions are involved in the transistor contrasted to the single junction in the diode. You might similarly suppose that the common-base configuration should be used because only one transistor junction would be involved in the load circuit. However, the basic physics of the device causes the saturated voltage drop between emitter and collector to be less than between base and collector. Thus, the common-emitter configuration is best for solid-state relays.

Two solid-state relays using common-emitter connected power transistors as contacts are shown in Fig. 5-8. Both make use of an optoisolator to separate input and output circuits. The relay of Fig. 5-8A operates in the normally open mode—without the proper input voltage, no load current can be passed. The relay of Fig. 5-8B simulates normally closed contacts—an appropriate dc input must be applied in order to interrupt load current.

In the circuit of Fig. 5-8A there is normally no forward bias applied to the base of the output transistor. If, however, a 5 V dc source is connected to the input terminals, the phototransistor will be conductive because of the radiation impinging on its photosensitive base. This will cause the normally off PNP transistor to conduct, thereby applying the necessary base bias to the output transistor to drive it into hard conduction. Thus, a voltage applied to the "solenoid" causes the relay "contacts" to close. Note that the phototransistor also has an electrical connection to its base region. In this circuit, the 10-megohm resistor between this base and the base of the output transistor helps speed up contact closure by regenerative action.

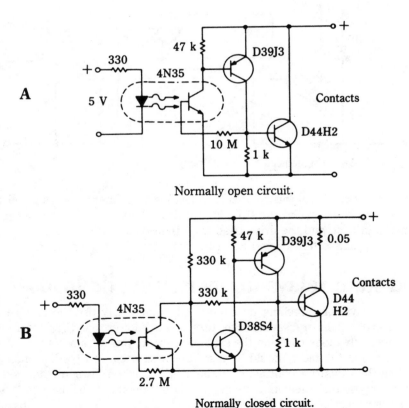

Normally open circuit.

Normally closed circuit.

5-8 Two 10 A, 25 V dc solid-state relays. General Electric Co.

(If a lower resistance were used, the circuit would behave as a latching relay—the output transistor would remain conductive after removal of the input voltage.)

In the normally closed circuit of Fig. 5-8B, an additional stage has been added to invert the result of activation of the LED emitter. The tiny resistance in the collector lead of the output transistor reflects the fact that maximum dissipation occurs during standby in this relay. This resistance drains off some power dissipation which could otherwise induce thermal runaway, particularly during overload.

In both circuits, the command situation at the dc input terminals can be negated by applying an appropriate bias voltage at the base lead of the phototransistor.

Interfacing a microprocessor with an ac load

Optical coupling between logic and power sections of a computer system is particularly appropriate, because it helps the overall system qualify for UL (Underwriters Laboratory) recognition. Practical examples will be shown for the Motorola TTL compatible microcomputer, the MC3870.

There are two ways to implement the power-control process—load power can be turned on either by a digital logic 0 or a logic 1. Different circuitries are required for the two situations. However, except for this transposition in functional logic, the operating principles of the two circuits are essentially the same. Both, for example, utilize a small transistor as a current booster for energizing the LED within the optoisolator. And, both switch the load power on or off by means of a triac.

The circuit for achieving logic 1 activation of the load is shown in Fig. 5-9. Transistor Q1 is needed because the MC3870 can deliver only about 300 μA

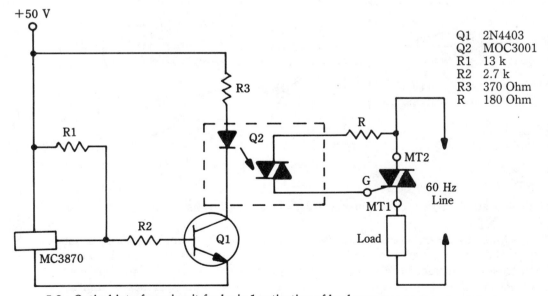

5-9. Optical interface circuit for logic 1 activation of load power. Motorola Semiconductor Products, Inc.

whereas the LED in the optoisolator needs 10 mA. The phototriac is triggered by the optical output from the LED, but otherwise behaves much like the triac trigger device commonly used in triac circuits. Thus, the load receives ac power through the main triac when the MC3870 outputs a logic 1. It should be noted that the firing of the main triac causes a near short to be placed across the phototriac, thereby turning it off; the main triac continues to conduct until commutated by the zero crossing of the ac line current. With the oncoming half cycle, retriggering of both the phototriac and the main triac occurs, providing the logic 1 level continues to be output by the MC3870.

Figure 5-10 depicts the circuit for achieving logic 0 activation of the load. Instead of the LED current being controlled by an NPN transistor, a PNP transistor is used. Because of this, the LED is energized when the MC3870 outputs a zero logic level. Otherwise, the action within the optoisolator and in the main triac load circuit is exactly the same as in the previously described interface scheme.

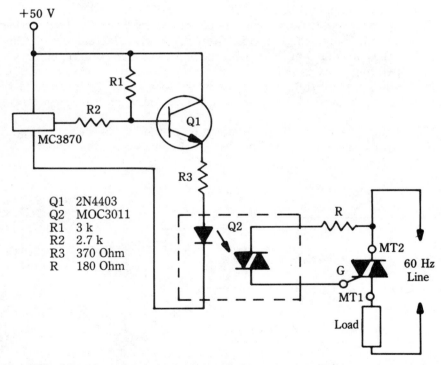

5-10   Optical interface circuit for logic 0 activation of load power. Motorola Semiconductor Products, Inc.

Other similar optoisolators can be used but heed must be paid to the LED ratings and to the peak blocking voltage. The optoisolators used in the above described applications are specifically designed as triac drivers. A group of such dedicated optoisolators listed in Table 5-1 enable appropriate values of limiting resistance R to be selected.

Table 5-1. Specifications of typical optically coupled triac drivers.

Device Type	Maximum Required LED Trigger Current (mA)	Peak Blocking Voltage	R(Ohms)
MOC3009	30	250	180
MOC3011	15	250	180
MOC3011	10	250	280
MOC3020	30	400	260
MOC3021	15	400	360
MOC3030	30	250	51
MOC3031	15	250	51

Motorola Semiconductor Products, Inc.

A group of suitable load triacs covering a wide current range is depicted in Table 5-2. These are nonsensitive gate types inasmuch as the optically coupled triac drivers can deliver 100 mA, and have one ampere peak ratings.

Table 5-2. Triacs for logic-interface systems with optically coupled drivers.

Triac	$T_{T(RMS)}$
MAC91, A	0.6
2N6068-75	4.0
2N6342-49	8.0
2N6342A-49A	12
MAC15, A Series	15
MAC223, A Series	25
2N6157-65	30
2N5441-46	40

Motorola Semiconductor Products, Inc.

Flashers

Flashers have many uses. They are widely applied as warning devices, visual alarms, and as attention getters. They serve many advertising purposes, and are widely found in toys, games, and novelty items. They are used in timing and in special illumination techniques, and they sometimes prove useful for instructional purposes. Most flashers either use filamentary lamps or LEDs, although fluorescent and gaseous discharge lamps can be used, too. The switching mechanism, once dependent upon mechanical devices or on various types of tubes, is now almost universally achieved with solid-state devices. The most popular format uses either one or two light-emitting elements; when two are used, they generally go through their on and off cycles alternately. Lenses and reflectors can be used, as the application dictates. Operating power can be derived from self-contained batteries, or from the ac utility line. Common denominator to most flashers is the use of some type of relaxation oscillator or multivibrator circuit; rate, and/or duty cycle can be fixed or adjustable. Sometimes, an audio output will be incorporated with the visual indication, as with certain alarm systems.

The most direct approach to flasher operation is to place visual indicators in the output leads of a conventional RC multivibrator circuit. Such a scheme is shown in Fig. 5-11, where LEDs indicate the conductive states of the two power MOSFETs. Do not use zener-protected MOSFETs for this application. The prescribed cautions must be exercised during handling and soldering in order to protect the gates from static and from power-line leakage and transients. Note that a constraint on the voltage swing at the gates is imposed by a positive-bias arrangement. There is more than meets the eye in this circuit. The almost infinite gate impedance of the MOSFETs can be exploited to produce very slow sequencing without the use of inordinately large coupling capacitors. However, it will then be found that the insulation quality of certain materials, such as phenolic boards, is inadequate. Indeed, the leakage in some printed-circuit boards might also be found excessive. The experimenter can readily modify this circuit to provide for the use of various filamentary lamps, inasmuch as many power MOSFETs will have adequate current capability for the purpose. The fact that this circuit does not need

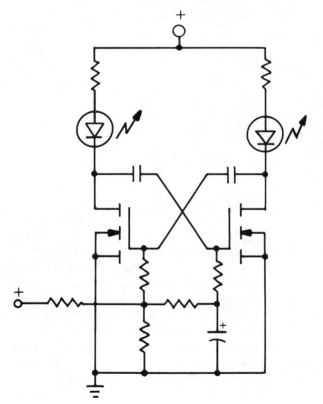

5-11   Flasher using power MOSFETs and LEDs in a multivibrator circuit.
Siliconix, Inc.

electrolytic coupling capacitors makes its operation more reliable and more pre-dictable than in an analogous design using bipolar transistors. Suitable power MOSFETs are the VN35AB, VN40AD, VN66AF, and VN88AD.

The 50-percent duty cycle that would normally accrue from this type of circuit can be altered by making the RC timing constants asymmetrical. Also, not only different-colored LEDs can be used, but entirely different types of lamps can be used for special applications. For more light output, several LEDs can be paral-leled in each drain circuit; however, it is then advisable to use separate current-lim-iting resistors in the individual LED circuits. And, although not commonly done, series-connected LEDs are also feasible with the advantage that only one limiting resistor is needed.

A battery-operated SCR flasher

The flasher shown in Fig. 5-12 is a proven design that has seen considerable use by manufacturers of warning devices. It is well suited for untended battery operation, and its parts count and cost are both low. It is not necessary to use the exact active devices designated in the circuit diagram, and there is considerable latitude in the selection of the battery and lamp. However, before attempting any modification of this basic circuit, a clear understanding of the operating principle should be attained.

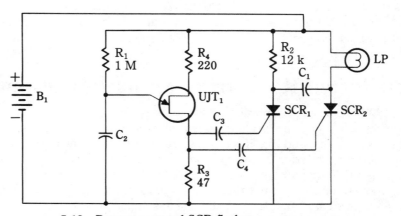

5-12 Battery-operated SCR flasher. International Rectifier Corp.

At first inspection, it might appear that the two SCRs form a multivibrator cir-cuit, with one of them (SCR2) providing the intermittent current for the lamp, and that the flash rate is synchronized to the frequency of the unijunction-transistor oscillator. This, however, is not the operational mode of the circuit. There is an SCR that is periodically turned on via a gate pulse from the unijunction-transistor oscillator, then subsequently turned off by a commutating pulse delivered from the other SCR. Thus, there is no operational symmetry between the SCRs—you could not, for example, put the lamp in the anode circuit of the other SCR, as would be the case with a multivibrator circuit.

It turns out that SCR1 is the commutating SCR; its purpose is to generate a negative pulse that passes through capacitor C1 and stops conduction in SCR2. A significant aspect of the mode of operation is that SCR1 never remains in its on state much beyond the duration of the narrow gate pulses received from the unijunction-transistor oscillator. This is because resistor R2 does not allow SCR1 sufficient holding current to remain turned on after termination of the gate pulse. This may be construed as the trick of this circuit. If it were not for this unique mode, circuit operation would quickly reach a dead end and there would be no repetitive on-off sequencing of the lamp.

Making use of the above information, the best way to visualize circuit operation is to assume the flasher is functioning as intended. Suppose your investigation commences at a time when both SCRs are turned off. A gate pulse from the oscillator then tends to turn both SCRs on. SCR2 turns on and remains on; SCR1 turns on only momentarily because of insufficient holding current from R2. It might be thought this momentary turn-on would commutate SCR2, turning it off. It is important to realize, however, that this does not occur. The turn-on action of SCR2 is stronger and takes precedence over the tendency of the commutating pulse from SCR1 to turn it off. The next gate pulse from the oscillator finds SCR1 in its off state and SCR2 already in its on state. Under this condition, the momentary triggering of SCR1 does commutate SCR2, turning it off. A subsequent gate pulse causes the cycle to start anew, with SCR2 again being turned on.

If an adjustable flash rate is desired, a variable resistor can be substituted for R1. With the circuit setup as in Fig. 5-12, about 60 flashes per minute will be produced.

A dedicated IC suitable for flasher systems

The LM3909 IC provides a novel function—it pulses an LED while operating from a single cell 1.5 V dc supply. The readily attainable pulse rate is in the vicinity of one flash per second, and this is easily modified upward or downward by appropriate selection of an external timing capacitor. The fact that the forward-voltage drop of LEDs is in the neighborhood of 1.7 V poses no obstacle, even for run-down flashlight cells. This is because the timing capacitor also participates in a voltage-incrementing action so that the output voltage can be 2 V or higher. Although it is true that this basic application does not involve enough power manipulation to qualify for consideration under the intended scope of this book, a quick look at its capabilities is likely to suggest incorporation into systems involving respectable power levels.

Several implementations of this interesting IC in flasher circuits are shown in Fig. 5-13. The single LED could just as well be the infrared-emitting diode of an optoisolator. In such an instance, drive can be derived for actuating powerful flashers via power transistors or thyristors. Figure 5-13A shows the basic LED flasher circuit. An alkaline D cell should last about two and one half years. A more useful foundation circuit is the variable flash-rate scheme shown in Fig. 5-13B. Here, the

A

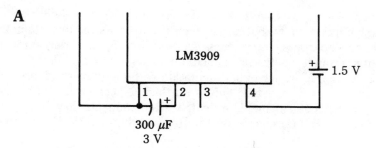

Basic LED flasher circuit. Nominal flash rate, 1 Hz.

B

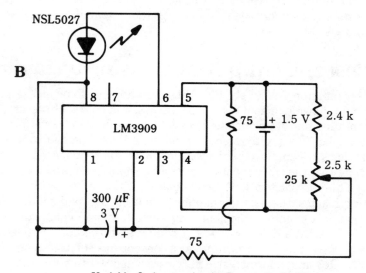

Variable flash-rate circuit. Range, 0–20 Hz.

C

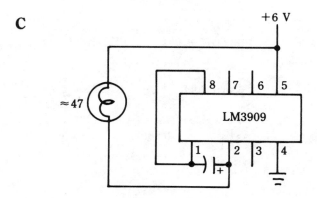

Incandescent-lamp flasher. Flash rate, 1.5 Hz.

5-13 Low-powered flashers using a dedicated IC. <small>National Semiconductor Corp.</small>

flash rate can be varied from 0 to 20 Hz. The incandescent-bulb flasher of Fig. 5-13C is included to show that higher output current is readily attainable. Thus, there should be no trouble interfacing this IC with an optoisolator, a transistor, or a thyristor for ultimate control of a moderately high-powered flasher system. By making use of this IC, much experimentation with multivibrators, oscillators, and timing elements can be avoided. This is worthwhile because it isn't always easy to deal with problems of leakage, inordinately sized elements, low input impedances of active devices, and unreliable startup that are common to circuits operating at flasher rates.

Known tolerance tantalum solid-state electrolytic capacitors are suitable for the timing function. Inexpensive garden-variety electrolytics of the common kind are satisfactory, but be prepared for capacity tolerances as sloppy as –20 to +100 percent. These devices might also exhibit too great a temperature dependency in some applications.

Incandescent lamp flasher using a thyristor

The incandescent lamp flasher shown in Fig. 5-14 appears to be an exercise in the use of thyristors, inasmuch as it uses diacs, SCRs, and triacs. However, its practical implementation and its principle of operation are quite straightforward. The heart of the flasher is a multivibrator circuit formed around the two SCRs. The two-terminal diacs are employed in conventional fashion for triggering the gates of the SCRs. Note that the SCRs operate from a dc supply comprised of a half-wave rectifier and a filter capacitor. This means that the SCRs must be forcibly commutated to their off states, that is, their anode-cathode current paths must experience an interruption or a negative transient long enough to stop conduction. This follows from the fact that a dc-operated SCR cannot be turned off by a gate signal.

Commutation of the SCRs occurs as a consequence of transients developed across resistors R3 and R4 when the SCR associated with one of these resistors

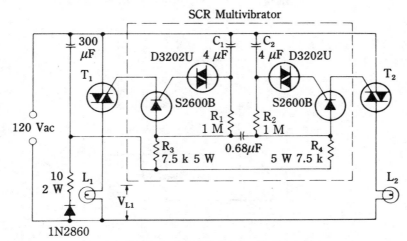

5-14 Thyristor multivibrator flasher. RCA

turns on. Assume, for example, that the left-hand SCR has just turned on; a large negative transient thereby generated across resistor R3 would then be communicated through the 68 μF capacitor to the anode of the right-hand SCR, turning it off.

The turn-on cycles of multivibrator action are governed by the time required for capacitors C1 and C2 to charge to the triggering voltage of their respective diacs; these, in turn, trigger the SCRs which serve as drivers for the lamp-controlling triacs. For example, if the right-hand SCR has just been commutated to its off state, capacitor C2 will charge through resistor R2 until the right-hand diac is triggered. Note that once the right-hand SCR is on, R2 is effectively connected to the negative side of the dc supply, thereby stopping the charging cycle of capacitor C2. At the same time, resistor R1 remains effectively connected to the positive side of the dc supply and charges capacitor C1 until the left-hand diac is triggered. As described above, this event abruptly turns off the right-hand SCR. These actions give rise to the alternate conductive state of the two SCRs, and to the corresponding flashing of the two triac-controlled lamps. Timing can be changed by altering the values of R1, R2, C1, or C2. R3 and R4 also affect timing, but these resistances should not be significantly modified because of their commutation function.

When selecting triacs, two ordinarily ignored characteristics of filamentary lamps must be taken into account. First, such lamps have high inrush currents when they are initially turned on because the resistance of the cold filament is much lower than that of the hot filament during normal operation. Inrush current for a 100 W lamp can be on the order of 10 A. Thus, a triac must have in addition to its continuous-current rating, an adequate peak-current rating. Unfortunately, ratings based upon inrush and operating currents only could prove woefully inadequate in the face of abnormal operation resulting from the burn-out of a lamp.

When a lamp burns out, you might suppose that a simple interruption of operating current occurs. The parting of the filament is accompanied by vaporization of the tungsten, giving rise to a momentary arc known as *flashover*. It imposes an extremely high current demand; in a 100 W lamp, the flashover current can be approximately 100 to 200 A. Taking into account the phenomena of inrush current and the shorter duration, but higher-amplitude flashover current, suitable 100 W lamp-control triacs (T1 and T2) for this flasher are the RCA MAC15-6 types. These triacs are rated for 15 A continuous loads and have 150 A single-cycle surge capability. (A single cycle at 60 Hz corresponds to 16.7 milliseconds, whereas the flashover duration of a 100 W lamp is generally less than 4 milliseconds.)

It might appear from the above the triacs for controlling 1000 W lamps in this flasher would have to have very high surge-current ratings, indeed. Actually, this is not the case; it turns out that the flasher current of 1000 W lamps remains about the same as for 100 W lamps. The RCA 2N5808 is a 40 A triac with a 300 A single-cycle surge rating. It is suitable for control of 1000 W lamps and its cost is much less than would be extrapolated from the situation involving the 100 W lamps.

Flasher using a PUT oscillator

The flasher shown in Fig. 5-15 uses a programmable unijunction transistor (PUT) as an oscillator, and is exceptionally easy to build and place into operation. The

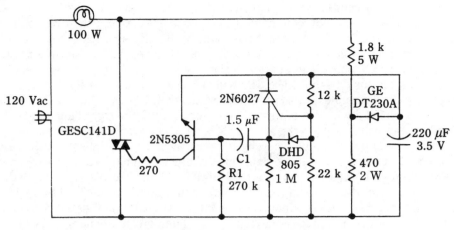

5-15 Flasher using a PUT oscillator. General Electric Co.

PUT is an anode-gate thyristor which lends itself particularly well to use in a flasher circuit. This is because a simple PUT relaxation oscillator allows more or less independent manipulation of repetition rate and duty cycle. The configuration of the PUT oscillator is slightly different from that commonly encountered in order to obtain a multivibrator-type waveform instead of sharp pulses. Although the internal operating mechanism is different from an ordinary unijunction transistor (UJT), the two devices develop similar negative-resistance characteristics. Like the UJT it is intended to replace in interest of greater design flexibility, the PUT is a small several-hundred milliwatt device.

In the circuit of Fig. 5-15, the triac output stage is interfaced with the PUT oscillator through the 2N5305 trigger circuit. Although not specifically drawn as such, the 2N5305 is actually a small Darlington transistor. The Darlington format imposes minimal loading on the oscillator circuit. When adapting this flasher to a particular situation, the flash rate will be found to depend primarily on capacitor C1. And, the flash duration can be adjusted by selection of resistor R1. Although the lamp triac circuit operates directly from the ac power line, the remainder of the flasher is dc powered. Direct current is provided by the simple half-wave rectifier circuit comprising the 1.8 k ohm and 470-ohm voltage-reducing resistors, the DT230A rectifier diode, and the 200 μF filter capacitor. In implementing this flasher, due consideration must be given to the lack of power-line isolation. (In this respect, it is similar to radios and TV sets.) A desirable safety feature would be a 1:1 power-line isolation transformer. Such a transformer, if not too large, can also provide current-limiting action to protect the triac from lamp flashover and from short circuits; by the same token, the lamp inrush current can be beneficially softened.

Stabilization of laser-diode optical output power

Many experiments and applications of laser technology are greatly facilitated by using injection laser diodes. These, for practical considerations, are similar to, but involve more sophisticated fabrication than, LED diodes. Indeed, the laser diode operates essentially as an LED at low currents; as current is raised, a threshold is reached at which the optical output changes mode from noncoherent to coherent radiation. In the later mode, the diode is said to be *lasing*. In the lasing region, current consumption and optical output power are highly temperature dependent. In order to protect the diode and to stabilize the optical output, special techniques are necessary. A much-used approach has been to maintain the temperature of the laser diode constant and operate it from a constant-current source. This involves a thermal-feedback system utilizing a temperature sensor, feedback, and a thermo-electric cooler. This is a workable scheme, but a simpler solution to the problem now exists.

Inspection of the stabilization circuit shown in Fig. 5-16 reveals that there are no thermal sensors, thermoelectric coolers, or constant-current sources. Instead, the laser diode has a built-in photodiode that monitors a tiny portion of the radiation.

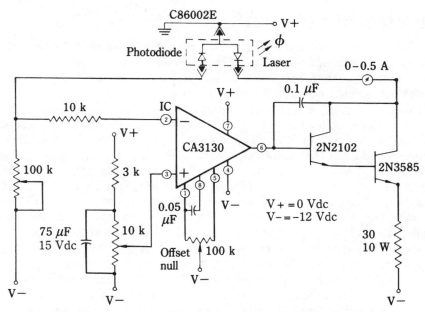

5-16   Laser diode system with flat output over a wide temperature range. RCA

This makes possible an optical-electronic feedback loop in which the current in the laser diode is automatically reduced if the radiation intensity tends to increase from a set level. Conversely, the current is increased when the radiation intensity tends to fall. The overall result is that the optical output is stabilized in the face of a wide range of temperature change. Note that the relationship between current and optical output is linear in the lasing region. Also, without such stabilization, it is possible for a sufficient temperature change to cause nonlasing operation.

In the circuit, the CA3130 op amp performs as a high-gain voltage comparator. It compares a reference voltage at its noninverting terminal, 3, with the voltage derived from the photodiode. The output (pin 6) of the op amp controls the conductivity of a Darlington transistor pair, which, in turn, controls the laser diode current.

A good quality series-pass voltage-regulated supply should be used to provide the 12–15 V operating dc for this stabilization system. Switching-type power supplies are not wise choices because laser diodes are susceptible to damage from transients.

WARNING

During operation, this laser diode produces invisible electromagnetic radiation that can be harmful to the human eye. This pertains to rays from reflecting surfaces as well as the direct beam.

An electronic siren

The electronic siren shown in Fig. 5-17 is probably best suited for use with toys, inasmuch as the output power level is in the one-third to one-half watt range. However, with subsequent power amplification, very useful siren systems can be implemented for vehicles, boats, and burglar alarms.

In order to simulate the sound characteristics of a mechanical siren, it is necessary to amplitude-modulate a tone by a lower frequency. From practical experience, the basic tone can have a nominal frequency of 1 kHz and the modulating frequency can be in the vicinity of several hertz. In Fig. 5-17, the power op amp portion of the LM389 is connected as a square-wave oscillator to supply the basic tone. Also, two of the on-board NPN transistors are connected in a multivibrator circuit with time constants such that oscillation in the several-hertz range can be obtained. The multivibrator interfaces the op amp oscillator via a third on-board transistor, which serves as the modulator. Modulation is accomplished because this transistor controls the dc current input to the op amp.

Both the basic-tone frequency and the modulation rate are manually controllable by variable resistances. A closer simulation to a mechanical siren would probably be obtained from the use of a sine-wave oscillator in place of the multivibrator. Also, an approach in this direction—gradual, rather than abrupt interruption of the basic tone—can be made by experimenting with RC networks in the base circuit of the modulating transistor. The circuit, as it stands, accomplishes the intended function of a siren, which is to capture the attention of the hearer.

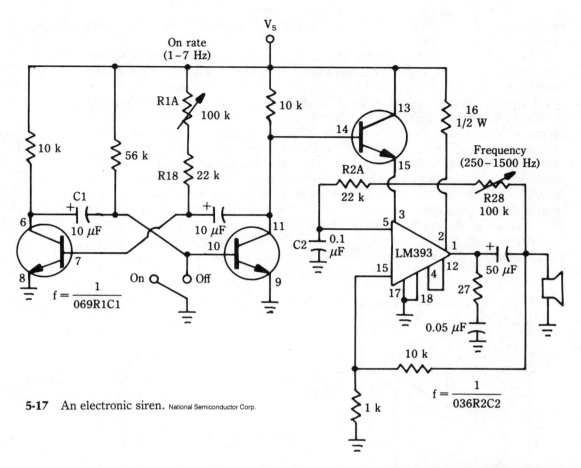

5-17 An electronic siren. National Semiconductor Corp.

Other op amps and other transistors can be used in carrying out the basic design of this circuit. The LM389 is, however, particularly suitable for such an application because of its low quiescent current drain, about 8 mA when operating from a 6 V supply. Optimum power output will result from using a 16-ohm speaker and a 12 V supply. A factor worthy of consideration in an electronic siren is the acoustic directivity of the speaker.

Power op amp
piezoelectric buzzer alarm system

Piezoelectric sound transducers are very efficient converters of electrical energy to acoustic energy. This is particularly true of those intended for use as alarms because they then operate at their self-resonant frequency. A practical approach to implementation of an alarm system utilizing the piezoelectric transducer is to place this device in the positive-feedback loop of an oscillator. If the active device of the oscillator has some power capability, all the elements of a powerful alarm system are then in place.

Shown in Fig. 5-18 is a piezoelectric alarm system configured around a programmable power op amp. The programmable feature enables it to be readily turned on or off from a logic source, if so desired. This op amp has a load-current capability of a quarter ampere. Although this capability is not directly made use of in this oscillator, such a rating puts it in a different league from the general run of op amps. Indeed, at higher dc operating voltages (the absolute maximum power-supply voltage for the LM13080 is 15 V), some transducers can be destroyed in this circuit. The intended transducer is the Gulton 101FB. Others can be experimented with, inasmuch as it has become standard practice to make these transducers with three terminals. The terminals are often designated drive, feedback, and ground. In this circuit, the ground terminal connects to the plus side of the dc supply. From an ac standpoint, this does not materially alter its mode of operation. The Archer Cat. No. 273-064 piezo buzzer, which operates at 6.5 kHz, is a good element for this circuit (sold in Radio Shack stores).

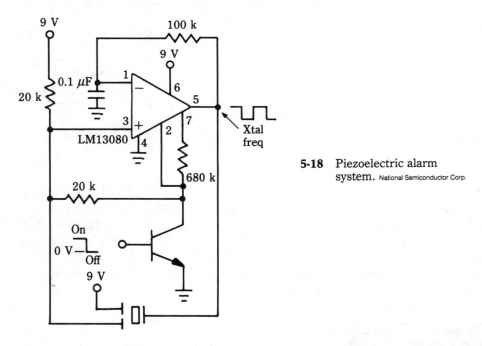

5-18   Piezoelectric alarm
system. National Semiconductor Corp.

Any of the common small-signal NPN transistors, such as the 2N2222, can be used as an electronic switch for enabling and inhibiting the alarm. This transistor controls the base bias in the input and output stages of the op amp. As a consequence of this type of control, dc power consumption is very low during inhibit. For best results, the output waveform at terminal 5 should show a 50-percent duty cycle. The duty cycle can be varied by experimenting with the 100 k ohm resistor, or possibly the 0.1 μF capacitor.

Interesting acoustic experiments can be conducted with this circuit. Piezoelectric elements operating at ultrasonic frequencies are also available. These higher frequencies have been used for animal control and insect repulsion. The output of these transducers is quite directional and often can be made even more so with small horns. Between 80 and 90 dB sound-pressure level is readily forthcoming. It might be possible to use two-terminal piezoelectric transducers by merely neglecting the ac ground connection. Such simple transducers can be made by simply forming a sandwich of two metal plates and a strip of appropriate barium-titanate material. This will be less efficient than a commercial buzzer, but facilitates experimentation in ultrasonics.

Power MOSFETs in AM transmitters

The RF power tubes in AM broadcasting stations have always posed reliability problems. They have limited life spans and, at best, slowly die in service from depleted emission. Also, they are bulky, expensive, difficult to cool, and require high-voltage power supplies. Solid-state design, until recently, has been unthinkable because of the multikilowatt power levels involved. However, practical designs have been demonstrated using large numbers of bipolar power transistors. A drawback to this scheme is that it requires about sixty of these power transistors per kilowatt. Because power transistors are physically small, it is feasible to use large numbers of them; the difficulty is that bipolar power transistors are not easy to parallel. Problems of current hogging and thermal runaway can be alleviated by the use of ballast resistors, but this seriously impacts operating efficiency and subtracts from the already meager power per unit that is available.

Contrarily, power MOSFETs are readily paralleled in large numbers, and do not normally exhibit thermal runaway characteristics. At the frequencies of the AM RF spectrum, these devices are just loafing along, whereas bipolar transistors incorporate undesirable parameter and ratings trade offs. Moreover, banks of power MOSFETs remain easy to drive, whereas bipolar types present awkwardly low input impedances when a large number of them are used. And, they do not need energy-wasting decoupling networks to prevent low-frequency oscillation— MOSFET gain is flat from dc to hundreds of megahertz. But, best of all, it only takes eight power MOSFETs to develop a kilowatt of output power.

A simplified circuit of the way in which power MOSFETs can be used to displace tubes in AM transmitters is shown in Fig. 5-19. Note that a bridge-configured RF power amplifier is provided with its operating dc by a switching power supply. By superimposing the audio signal on the error voltage, the output of the switching supply is audio-modulated about its mean voltage level. This, of course, directly modulates the RF power amplifier. An interesting aspect of this scheme is that the IRF350 power MOSFETs suggested by International Rectifier Corporation are not packaged as RF types. The implication is that the frequency capability of these devices is so good that no special techniques are necessary at AM broadcast frequencies.

The IRF350 carries maximum ratings of 400 V and 15 A. The allowable power dissipation at 25°C is 150 W. Here again, advantages in practical implementation

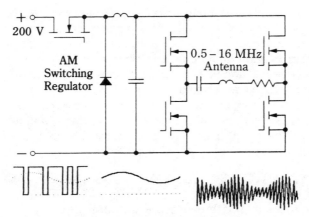

5-19 Simplified circuit of a power MOSFET AM transmitter. International Rectifier Corp.

are realized over bipolar transistors, namely that operating impedances tend to be more tubelike. Finally, the absence of the secondary-breakdown phenomenon inherent in bipolar transistors, frees the power MOSFETs from catastrophic breakdown during modulation peaks.

Modulation scheme for power-MOSFET RF output amplifiers

Another unique method of modulating power MOSFETs is shown in Fig. 5-20. Probably more experimental in nature than the high-level modulation scheme used in Fig. 5-19, this approach is analogous to the grid or efficiency modulation formats long used in tube transmitters. A profound difference from the tube designs is that modulation is accomplished via a PWM wave. Note that a switching power supply is not needed in this case; rather, the pulse-width modulated wave is impressed, along with the incoming RF drive, at the gate of the power MOSFET. (Actually, a number of power MOSFETs can be paralleled in order to attain a higher power output.)

A basic requirement of this scheme is that the triangular waves must have a frequency at least several times higher than the cutoff frequency of the low-pass filter in the output circuit of the amplifier. This is so in order to enable the filter to effectively prevent the PWM interruptions from appearing in the envelope of the amplitude-modulated waveform. From a practical standpoint, the low-pass filter must first satisfy the tank circuit and impedance-matching needs of the RF amplifier. Then, a sufficiently high level PWM wave is impressed at the gate to produce a clean modulation envelope at the output. With low-frequency transmitters, this can be readily accomplished, but it becomes increasingly more difficult to do so beyond transmitter frequencies of several megahertz or so. It then becomes more difficult to produce the requisite high-frequency PWM format, and contradictions in the design of the low-pass filter tend to become more aggravated.

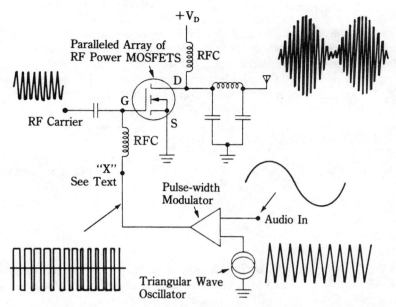

5-20 AM modulation technique for MOSFET transmitter or final amplifier.

A modification to the arrangement shown in Fig. 5-20 can resolve difficulties such as those described. By inserting, at point X, a separate low-pass filter with a cutoff frequency just above the audio spectrum of interest, the gate will be impressed with an audio signal rather than the chopped PWM wave. The situation will then simulate the grid-modulation technique used in tube transmitters. The output low-pass filter, or pi network, in the drain circuit of the power MOSFET can then be optimized entirely for maximum performance as a tank circuit and impedance matcher. This also relaxes the high-frequency requirements of the triangular wave; it still must be much higher than the highest audio-frequency, but no longer bears any relationship to the transmitter carrier frequency.

Here again, power MOSFETs can be readily paralleled; this is particularly beneficial in low-level, or efficiency modulation, where RF output power is about one fourth that attainable with high-level modulation. (For power MOSFETs, high-level modulation results from modulating the drain circuit, as is done in Fig. 5-19.)

Voltage-regulator IC
as an amplitude modulator

The three-terminal IC voltage regulator is sometimes found in other than its intended application as a simple, compact regulated dc supply. This makes practical sense considering this ICs sophisticated amplifying circuits, stable reference voltage, power-handling capability, thermal protection, and all at low cost. Lambda Semiconductors suggests the unique amplitude-modulation scheme shown in Fig.

5-21. This, or a similar arrangement could conceivably be used to provide AM (amplitude modulation) for a transmitter. (AM continues to be widely used in VHF (very high frequency) aircraft communications, and in other services.) Unity or relatively low-voltage amplification is obtained, but current and power gain are very high. Frequency response is no problem inasmuch as these ICs can be expected to display flat response to something on the order of 100 kHz or so.

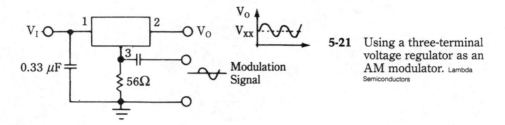

5-21 Using a three-terminal voltage regulator as an AM modulator. Lambda Semiconductors

Best results will be forthcoming with adjustable voltage regulators, but many fixed types can be anticipated to also work well when used in this fashion. Lambda Semiconductors seems to feel this operational mode should be restricted to ICs with maximum output-current ratings of 1 A. The experimentally inclined will, no doubt, be able to devise ways of safely scaling up current, if needed. It might well be that paralleling is more easily accomplished in this mode than it is in conventional voltage-regulator applications. And, liberal heat sinking might also prove profitable. In any event, this alternate use for the voltage-regulator IC is an intriguing one; comparing it to conventional modulator designs, it appears likely that a reduction in parts count, cost, and developmental effort can be achieved.

In many three-terminal IC voltage regulators, terminal 1 (in Fig. 5-21) is the input terminal, and terminal 2 is the output terminal. Terminal 3 is generally designated as ground in fixed-voltage types, but as adjust in adjustable types. The coupling capacitor for the audio signal should be about 1000 μF. Judging from transmitter designs using all discrete devices, it is likely that deep modulation of a class C RF power amplifier can be accomplished without need of a modulation transformer. (In order to increase modulation percentage, it might be necessary, as is standard practice, to apply some modulation to the driver along with the final amplifier.) In addition to RF transmitters, other applications in instrumentation and in hobbyist activities will, no doubt, suggest themselves. Conceptually, the direct modulation of power-supply voltage is a more elegant design technique than the traditional method of placing a separate modulator in series with the dc source.

Power MOSFET in horizontal-sweep circuits

In the television-servicing industry it has become common knowledge that catastrophic failures are very likely to involve the horizontal output stage. Both power transistors and thyristors have been used in solid-state TV circuitry, and both have produced their share of failures. Such failure often destroys other devices and components so that it is not always entirely clear which was the cart, and which the horse. Nonetheless, enough experience has been gained to suggest that reliability would be enhanced if the vulnerability to failure of the horizontal output stage could be reduced. This applies not only to home TV sets, but to other cathode-ray tube display equipment as well.

Many reasons could be advocated to account for this breakdown phenomenon. Obviously the margin of safety designed into the set does not always encompass the simultaneous occurrence of worst-possible operating conditions. Transients, elevated temperatures, thermal cycling, aging, and the effects of both low- and high-line voltage all contribute to reduction of mean time between failure, but the combination is not easy to quantify. The situation is frustrating to the public, and to professionals as well, because (right or wrong) it was widely thought that solid-state equipment would last forever. Moreover, when a solid-state horizontal output device self-destructs, the repair is far more involved, and is costlier than was the simple replacement of a horizontal output tube that had gradually become gassy or lost emission.

Although transistors and thyristors with higher ratings would, no doubt, help matters, this is easier said than done, especially when cost is considered. A possible solution, however, is the use of the power MOSFET. Not only is excellent high-speed performance easy to attain, but considerable circuit simplification results. These devices are easy to drive; the usual power-driver stage and transformer are no longer needed. Best of all, the nature of the SOA curve of these devices is such that failure is less likely. The forgiving nature of the power MOSFET stems partially from the simple silicon output element which involves no PN junction. Then, too, the temperature coefficients are opposite those of PN junction devices which tend to go into thermal runaway if overloaded. The power MOSFET, conversely, tends to limit its current consumption. The reason designers haven't hitherto used power MOSFETs in horizontal output stages is because devices with the requisite rating have not been available. This is no longer true and it appears likely that this device will significantly contribute to the reliability of cathode-ray tube equipment.

The power MOSFET horizontal-sweep circuit shown in Fig. 5-22 illustrates the lack of clutter generally attending bipolar-transistor and SCR circuits. The turn-off time of the power MOSFET is about 150 nanoseconds compared to 3.5 microseconds or so for the bipolar transistors used for such service. This has many favorable ramifications pertaining both to operation and reliability. The reason for the speedy turn off is that majority carriers only are involved in MOSFET operation; the minority carriers that produce the undesirable charge-storage effects in PN junction devices have negligible involvement in MOSFETs.

Conspicuous by its absence in the simple circuit of Fig. 5-22 is the phase-locked loop often needed in bipolar-transistor horizontal-sweep systems to compensate for timing errors due to charge-storage phenomena. The requirement for such corrective circuitry stems from the fact that such charge storage varies from transistor to transistor and is a complicated function of temperature and operating conditions. The ability to dispense with such extra-baggage circuitry further enhances the operational stability and the reliability of the power MOSFET horizontal-sweep system.

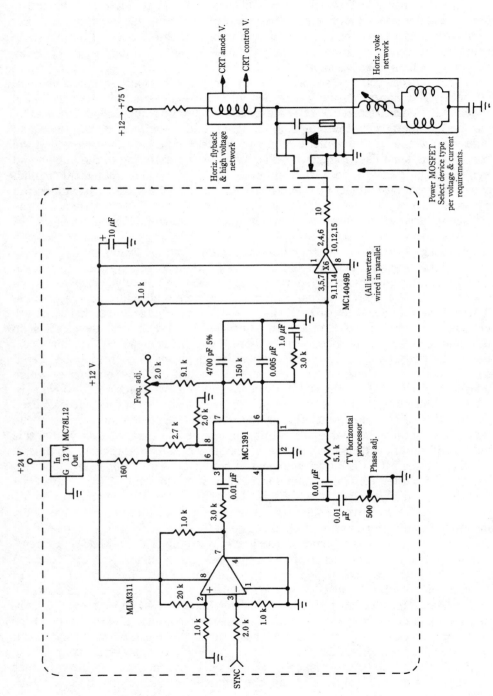

5-22 A power MOSFET horizontal sweep circuit. Motorola Semiconductor Products, Inc.

The drive for the power MOSFET is provided by six logic-circuit inverters connected in parallel; these are contained within a single MOS IC. The rationale underlying this drive technique is to provide a speedy discharge of the gate source input capacitance of the power MOSFET at turn-off time. This type of storage is easy enough to deal with. In contrast, the minority-charge storage in the base-emitter region of bipolar transistors cannot be driven out fast enough to yield a shorter turn-off time than the mentioned several microseconds.

An alternative to the MC1391 horizontal-processor IC is the SGS TDA1180; there are others too, but not necessarily with pin compatibility to the MC1391. The common denominator of these dedicated ICs is that they contain a phase detector and a voltage-controlled oscillator. At this writing, none of these have direct-drive capability for a power MOSFET output stage. Looking at the overall circuit of Fig. 5-22, it is apparent that an improved horizontal-processor IC could conceivably contain the sync inverter and the power MOSFET driver. And, judging from what has been accomplished in the power IC field, it appears reasonable to suggest that the 12 V regulated-dc supply could also be incorporated within the proposed module. What appears to lie immediately ahead are extremely simple horizontal-sweep systems with stable and precise performance, and characterized by very high reliability. Also, much of what has been said also applies to vertical-sweep systems, although electrical stresses tend to be less than in horizontal-sweep systems.

A two-transformer 100 W ultrasonic inverter

High-frequency inverters are useful as building blocks of regulated power supplies, and as the ac power source for fluorescent lighting, ultrasonic transducers, and specialized welding techniques. Such inverters can be readily implemented as saturable-core oscillators and often have a single transformer that operates at the output-power level of the inverter. However, with a slightly increased parts count, the inverter can be made to operate more efficiently and more reliably.

The single-transformer saturable-core inverter necessarily sustains high losses from magnetic hysteresis. It also tends to be temperamental with regard to its loading and starting behavior; refusal to start when loaded is an often-encountered phenomenon. Although not always of any consequence, the single-transformer inverter is relatively vulnerable to frequency change as a function of loading. These shortcomings can be minimized via various design techniques, but it is probable that the two-transformer inverter is inherently a better performer. It operates on essentially the same principle as the single-transformer inverter, but a small saturable-core transformer is used in the input circuit. A large transformer is used in the output circuit, but it operates in its linear mode—it is not allowed to saturate.

A practical two-transformer ultrasonic inverter is shown in Fig. 5-23. It delivers 100 W of square-wave power at an adjustable frequency from 25 to 40 kHz. The nominal output voltage is 140 V, but this can be readily changed to suit unique needs by altering the secondary turns on the output transformer. The operating efficiency is about 93 percent. It will be noted that this inverter is operated directly

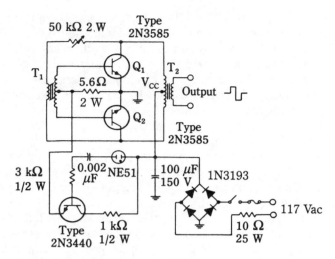

T₁ = Allen Bradley RO-3 (EI 102 H 142 A) or equiv.
   primary: 160-turn #32 wire;
   secondary: each 3-turns #32 wire.
T₂ = Indiana General C2 material (CF216) or equiv.
   primary and secondary: 80-turns #28 wire.

5-23 A two-transformer, 100 W ultrasonic inverter. RCA

from rectified line voltage—there is no power transformer associated with the rectifier. This considerably reduces weight, bulk, and cost. It also makes overall efficiency higher than specifications usually indicate for inverters operating at lower voltages.

Associated with the basic two-transformer inverter is a starting circuit. This auxiliary circuit gives the inverter transistors a jolt of forward bias when the power-line switch is closed. There is no danger of sustained simultaneous conduction of the inverter transistors. This is because of the transient nature of this starting bias. Also, the slightest difference in gain or speed between the two transistors in the inverter quickly causes one of them to take over and initiate normal oscillation.

Frequency is adjustable by the 50 k ohm variable resistor in the feedback loop. Although topographically asymmetrical, the circuit is electrically balanced and a good 50-percent duty-cycle wave is obtained over the entire frequency range.

Capacitor-discharge ignition system

By the mid-seventies, most automobiles had adopted electronic ignition. Although less standardized than the long enduring breaker-point ignition, the various circuits and formats all yielded beneficial performance features. These included better gas mileage, less susceptibility to fouled plugs, easier starting—particularly in cold weather and with low-battery capability—less tendency to drop out at high engine speeds, and reduced maintenance. There remain, however, many older cars on the road with the more primitive ignition systems. These cars have not been displaced for a variety of reasons. Nostalgia, economics, and user satisfaction are obvious enough reasons for retaining these cars. Many people view them as higher-quality products than their more modern descendents. And, certainly, many feel safer in a heavily constructed automobile than in a tiny minicar, fuel mileage notwithstanding. Whatever the reason, most of these older cars can benefit from substitution of an electronic ignition system.

The most convenient type of electronic ignition system for this purpose is one that does not dispense with the breaker and condenser within the distributor. This might initially appear as a contradiction inasmuch as one of the declared advantages of electronic ignition is that the breaker points can be eliminated, and with them a lot of headaches and maintenance. However an electronic ignition system that retains the use of the breaker points still drastically reduces maintenance associated with them. This is because the current in the breaker points is, so low that there is virtually no arcing, burning, or sparking. Moreover, the clearance between the points is less critical with electronic ignition. The nice thing about retaining the breaker points is that it is then easy at any time to revert back to the original ignition system. Indeed, some of the after-market systems have a switch just for this purpose.

Of the different types of electronic ignitions, the capacitor-discharge type is particularly well suited for upgrading older automobiles. This type of ignition system readily provides a steep wavefront, high-energy pulse that is effective in firing fouled sparkplugs. And it is no trick to design such an electronic ignition system so that, in addition to the breaker-points and condenser, the old ignition coil can also be retained and used. Many of these older cars are not too easy to work on, so the less installation surgery, the better. Some car buffs, to be sure, go all out and install magnetic reluctance or optical-pickup breaker systems and procure special ignition coils. Most owners will not wish to cope with the extra effort and cost of such refinements because the added improvement becomes marginal in most older cars.

The direct approach to capacitor discharge ignition is to use an SCR to discharge a capacitor through the primary of the ignition coil. There, of course, has to be some means of commutating the SCR inasmuch as dc-operated SCRs ordinarily remain in their conductive state once triggered on. There also has to be a means of quickly charging the storage capacitor to a potential of several hundred volts in order to give the system the requisite energy level. (Incidentally, when ignition breaker points open in the old-style ignition system, the voltage across the primary of the ignition coil is not the nominal 13.5 V from the battery, but is several hundred volts because of the counter EMF from the coil.)

Such an electronic-ignition system generally requires auxiliary circuitry to help stabilize performance over a wide range of conditions—at one extreme, there is the cranking of a cold engine with low battery voltage; at the other extreme, battery voltage might be at or near its maximum while a fast-running engine demands fast but energetic firing of its plugs. The simplified block diagram of Fig. 5-24 shows the format for a typical capacitor-discharge ignition system using an SCR. (The fact that ignition coils are usually three-terminal autotransformers rather than the shown four-terminal format is of no consequence in demonstrating the

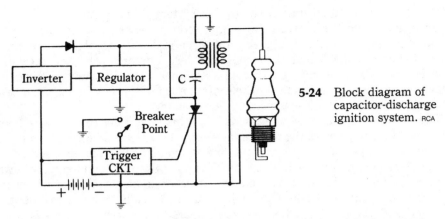

5-24 Block diagram of capacitor-discharge ignition system. RCA

basic principles involved.) Between firings, high voltage from the inverter (or more precisely, the converter) charges the storage capacitor; when a plug is to be fired, the SCR is triggered to its on state, thereby dumping the accumulated charge into the primary of the ignition coil. Then the SCR must be quickly turned off in preparation of the next charging cycle. This is straightforward enough, but keep in mind that this cycle of events is repetitive at a very rapid rate in a fast running multicylinder automotive engine.

The circuit of a practical capacitor-discharge ignition system is shown in Fig. 5-25. There are two phases to the construction of this system, the procurement, mounting, and connecting of the components, and the construction of the inverter transformer, T1. The inverter, together with the 1N3195 rectifier diode, becomes a dc-dc converter. Q1 is the switching transistor, and Q2 is its driver in the fairly conventional flyback inverter circuit. Q3 is a control stage; in conjunction with Q4, it provides some of the auxiliary functions alluded to. The first step in building the capacitor-discharge ignition system is to construct the inverter transformer. Table 5-3 lists the parts needed and specifies the important data for making the transformer. Figure 5-26 shows how the individual windings are connected. In this type of application, there is some latitude in selection of the dimensions of the E-I segments used in the core of the transformer.

A more detailed insight into the operation of the circuit is acquired by considering the functions of transistor stages Q3 and Q4. Because of its circuitry position, Q3 has the capability of either shutting down or limiting the amplitude of the flyback pulse delivered to storage capacitor C2. Both of these actions are advantageously used. When the ignition breaker points are open, Q3 completely shuts down the inverter circuit. It does this via sensing resistor R7; when the breaker points are open, the base of Q3 is sufficiently positive to cause heavy conduction in Q3, thereby depriving drive-stage Q2 of operating bias. It follows that the flyback output transistor, Q1, is also inhibited from normal operation. If it were not for this provision, the inverter would have to operate into a short-circuit during those times when plugs were being fired.

Note also that the base circuit of Q3 senses the voltage developed across storage capacitor C2. It does this through R8 and the 12-volt zener diode. Because of

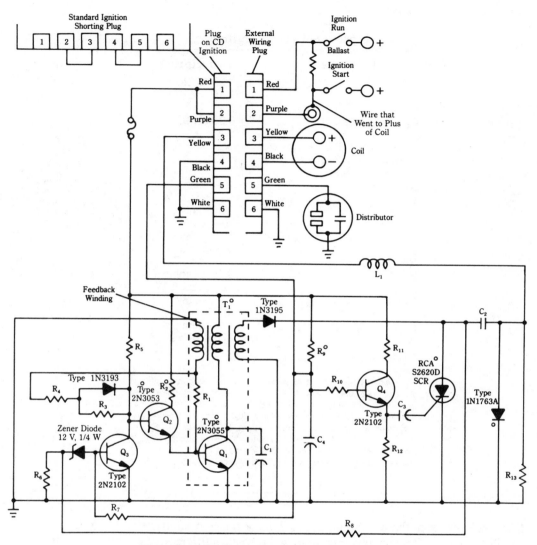

5-25 Capacitor-discharge ignition system. RCA

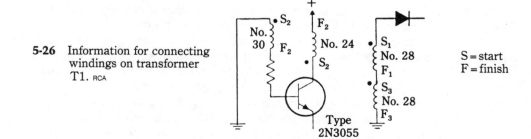

5-26 Information for connecting windings on transformer T1. RCA

S = start
F = finish

**Table 5-3. Parts list for
capacitor-discharge ignition.**

Part	Description
C_1	0.25 μF, 200 V
C_2	1 μF, 400 V
C_3	1 μF, 25 V
C_4	0.25 μF, 25 V
F	5 A
L_1 10 μH, 100T of No. 28 wire wound on a 2 W resistor (100 ohms or more)	
R_1	1000 ohms
R_2	50 ohms, 5 W
R_3	22,000 ohms
R_4	1000 ohms
R_5	10,000 ohms
R_6	15,000 ohms
R_7	8200 ohms
R_8	0.39 megohm
R_9	220 ohms, 1 W
R_{10}	1000 ohms
R_{11}	68 ohms
R_{12}	4700 ohms
R_{13}	27000 ohms

All resistors are 1/2 W unless otherwise indicated

T_1 = Transformer, wound as follows: A 1/2 in. bobbin and El stack of grain-oriented silicon steel are used; first, 150 turns of No. 28 wire are wound and labeled start 1 and finish 1 on the winding; second, 50 turns of No. 24 and No. 30 wires are wound bifilar and labeled start 2 and finish 2; third, 150 turns of No. 28 wire are wound and labeled start 3 and finish 3. All windings are wound in the same direction. A total air gap of 70 mil (35 mil spacer) is used. Connections are made as shown in Fig. 5-26.

RCA

this feedback loop, the amplitude of the flyback pulse is regulated; when the voltage in capacitor C2 tends to get too high, Q3 is made sufficiently conductive to partially shut down the inverter by inhibiting its natural regenerative activity. If it were not for this technique, the voltage delivered to the distributor and spark plugs could get high enough to endanger insulation. (Approximately 350 peak volts across C2 corresponds to 40 kV open circuit at the spark plugs.)

Q4 and associated circuitry provide the trigger signal to the gate of the SCR when the breaker points open. This stage effectively isolates the SCR from stray and transient voltages that might otherwise trigger it. Examples of these false gate signals are those due to breaker-point bounce immediately after closing, residual voltages across the closed breaker points because of imperfect contact, and the inverter frequency riding on the battery supply leads. Because of capacitor C4, the original capacitor across the breaker points can be removed; however, it is desirable to retain it. Not only does it provide additional suppression of spurious voltages, but its intention enables quick reconversion to the original ignition system in an emergency, or during the course of testing and evaluation.

Those items in the schematic diagram which are designated with a small circle should be accorded special heat-removal treatment. Solid-state devices tending to run hot should be heat sinked. These include the SCR, the 1N1763A commutating diode, and the two inverter transistors, Q1 and Q2. Provide as much convection as possible for transformer T1 and secure power-dissipating resistors to aluminum or copper surfaces with heat-conductive epoxy. Note that the original ballast resistance is automatically shorted out when the new ignition system is plugged in. This is permissible because of the internal electronic regulation.

A simple 12 V battery charger

A 4 A battery charger is a practical and convenient adjunct for vehicles and boats with 12 V batteries. Usually, a battery has been depleted because the lights were left on, or because of excessive use of the starter under difficult starting conditions. In any event, such dead batteries can often be brought up sufficiently for starting purposes within a few hours when charged at a 4 A rate. At worst, overnight charging is bound to suffice. Although quicker results can be forthcoming at higher charging rates, there is ample evidence that this shortens the lifespan of the battery. Also, higher charging rates tend to produce excessive gassing, and this is sometimes dangerous.

A good design approach for such a charger is a basic series-pass dc voltage regulator with a current-limiting circuit to prevent the charging rate from exceeding 4.5 A. Such a circuit is shown in Fig. 5-27. It uses a power MOSFET as the series-pass element, and an op amp as the error amplifier. This works out nicely

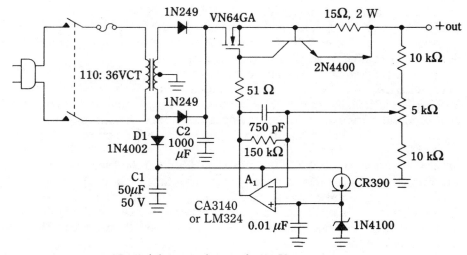

5-27 4 A battery charger for 12 V systems. Siliconix, Inc.

because of the meager drive requirement of the MOSFET—if a bipolar transistor were used, at least one additional driver stage would be required. Current limiting occurs when the 2N4400 NPN transistor becomes forward-biased from passage of the charging current through the 0.15-ohm sampling resistor; the 2N4400 then conducts, thereby depriving the power MOSFET of any additional drive voltage.

The error amplifier compares the zener-developed reference voltage with an adjustable sample of the output voltage, and varies the drive to the power MOSFET in such a way as to null the error voltage and maintain the output voltage constant. Note that the zener diode is supplied by a constant-current diode, rather than by the usually used resistor. This enhances the stability of the reference voltage, and also simplifies certain design problems.

Even though the primary of the transformer circuit is fused, and output current is limited to 4.5 A, it would be wise to equip the MOSFET with a liberal heat sink. This will protect it from excessive temperature rise that could result from inadvertent short circuit of the output—a common occurrence under the unfavorable working conditions often accompanying battery failure. Also, although not shown a 0–10 A ammeter will make the application of the charger more meaningful. This is particularly true because some feeling of charging progress can then be obtained—a good battery will accept less charging current as it approaches the fully charged condition.

Burst-modulation proportional control of a heating system

Burst modulation, otherwise known as integral-cycle control, is particularly well suited for the control of temperature in heating systems. The inherently slow rate of change in such a power system enables very refined feedback control in terms of the number of power-line cycles the load is permitted to be turned on at any given time. Such control is readily made proportional in the sense that the greater the required temperature rise, the greater the number of cycles of power to the heater. This contrasts to the overshoot-undershoot temperature control commonly provided in households via thermostatic switches.

From the electrical standpoint, burst modulation generates negligible RFI or other electrical noise because power is initiated and terminated at zero-voltage crossings. Additionally, such control subjects the control thyristor to considerably less electrical stress, thereby enhancing system reliability. And when used for large-lamp control, burst modulation definitely extends filament life because of reduced inrush current.

The circuit shown in Fig. 5-28 is designed around the Motorola MFC8070, a zero-voltage switch capable of directly triggering large triacs. This IC will either deliver gate-trigger pulses for each half cycle of power-line frequency, or will be in a quiescent state. Which of these two states prevails is dependent upon the relative logic levels at two inputs (pins 2 and 3) of the device. In any event, the trigger pulses occur only at zero-voltage crossings.

The other two function blocks actually comprise a single IC, a dual D flip-flop,

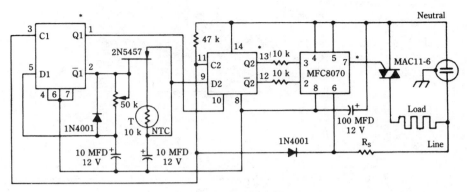

*McMOS Dual D Flip-Flop MC14013L

Line voltage	R_S
120 Vac	10 k 2 W
240 Vac	20 k 4 W

5-28 Burst-modulation circuit for proportional control of temperature.
Motorola Semiconductor Products, Inc.

the Motorola MC14013L. This IC is made with MOS (metal-oxide semiconductor) logic. The two flip-flops are used to initiate and terminate voltage ramps in two RC circuits and to inform the zero-voltage switch when to turn off the thyristor gate pulses. Significantly, one of the two RC circuits contains a thermistor—its charging rate is therefore dependent upon the temperature sensed by the thermistor. The other RC circuit contains a 50 k ohm variable resistor by which its charging time can be manually adjusted. Note that both flip-flops receive line-frequency clock pulses from half-wave rectification via the R_S-1N4001 circuit.

The above has defined the functions of the important components in this system. To more clearly see how they work together, refer to the circuit and to the waveform diagram of Fig. 5-29.

Beginning at the left part of the waveform diagram, note the initial state of flip-flop #1 is such that $\overline{Q1}$ is high and $\overline{Q2}$ is low. Note that the voltage level at D1 is rising. The first significant event is the attainment of threshold level by the charging voltage at D1; this reverses the logic state of flip-flop #1, making $\overline{Q1}$ low. Referring back to the circuit, appreciate that as long as $\overline{Q1}$ is high, both RC circuits are being charged, that is, their respective 10 μF capacitors are developing voltage ramps. However, when $\overline{Q1}$ goes low, both RC circuits are discharged—one through a 1N4001 diode, and the other through the gate junction of the 2N5457 JFET.

The discharge of the RC circuit associated with flip-flop #2 is sensed at D2 and causes a reversal of logic state in flip-flop #2. Thus, the next significant event is the transition from low to high level at $\overline{Q2}$. Because the zero-voltage switch now senses a reversal of logic levels at its input, it is enabled to generate gate trigger pulses for the triac.

A subtle, but very important event in the sequence should now be recognized. When $\overline{Q1}$ became low, it discharged both RC circuits. Accordingly, at the next

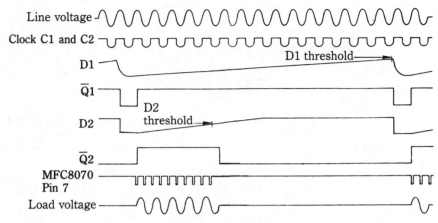

5-29 Waveforms for circuit of Fig. 5-28. Motorola Semiconductor Products, Inc.

clock pulse, D1, being deprived of this threshold voltage level, makes flip-flop #1 again reverse its state. Thus $\overline{Q1}$ reverts to logic high and starts a new charging cycle for the voltage ramps at D1 and D2.

Proceeding to the right on the waveform diagram, the next significant event is the attainment of threshold voltage level at D2, changing the state of flip-flop #2 so that $\overline{Q2}$ is again at logic low. This event restores the zero-voltage switch to its inhibit state, and power to the load is turned off. After the elapse of more time, the voltage at D1 attains its threshold level and the initial events then repeat. Thus, the extreme right-hand portion of the waveform diagram resembles the extreme left-hand portion where the analysis commenced.

Note that the duration of applied power to the heater and the length of time between such applications are both dependent upon charging rates. A low ambient temperature (cold thermistor) slows the charging rate at D2, thereby delaying termination of heater power. And increasing the 50 k ohm variable resistor (corresponding to a lower temperature requirement) slows the charging rate at D1, thereby delaying initiation of heater power. In a well-engineered heating system, the time between wave trains of power would not be too long and a person would experience smaller increments of temperature change than is usually the case with the commonly used thermostatic-switch type of control.

A PWM audio amplifier

Not too long ago, hi-fi enthusiasts became understandably excited over a unique concept of audio power amplification. The technique promised very high efficiency, a clean solution to thermal problems, and worthwhile reduction in the space and weight required per watt. Moreover, it was projected that before long, cost per watt would also be lower than with conventional power amplification. Although performance was not as good as with conventional methods, it was demonstrably acceptable, and there appeared no theoretical reason why it could not be expected to be improved with time. Many prototypes, and even a limited number of commercial models, convincingly showed the scheme to be practically feasible—or almost so. As might be expected with new models in almost any technology, there were, admittedly, some bugs to be contended with. For a number of reasons, and not all of then technical, the off-beat audio system did not survive the setbacks encountered in the introductory models. The basic idea has remained in the contemplations of engineers, and experimenters feel that practical implementation will again see the light of day.

The technique referred to is class D power amplification, otherwise known as the pulse-width modulation amplifier. Although it did not successfully invade the domain of hi-fi and stereo in its first attempt, this technique has enjoyed widespread use in regulated power supplies, and to a lesser extent in servo systems involving the control of motors. A great deal has been learned, and many improved and dedicated solid-state devices have become available to implement such systems. It is not farfetched to suppose that renewed investigation of the possibilities of this approach to audio power may produce fruitful results.

The principle of the PWM audio amplifier is surprisingly simple. As seen in Fig. 5-30, a pulse-width modulated format is generated when an overdriven summing amplifier is simultaneously impressed with the audio signal and a high-frequency triangular wave, known as the carrier. The class D amplifiers are then turned on and off by the pulse-width modulated wave train. Here is where efficiency enters the picture; because the class D amplifiers are either on or off but never in an in-between state, their theoretical efficiency is 100 percent. (This, of course, is based on the ideal premise that their on resistance is zero, that they make instantaneous switching transitions, and that their off resistance is infinite.) The output of the class D amplifiers is passed through an integrator in the form of

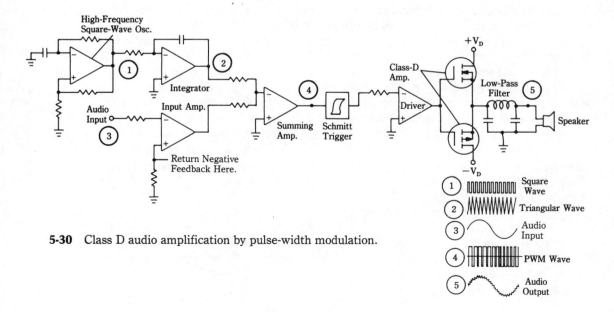

5-30   Class D audio amplification by pulse-width modulation.

a low-pass filter with a cutoff frequency above the audio range, but much lower than the carrier frequency. What emerges is the restored audio wave, stripped of the interfering carrier frequency.

In Fig. 5-30, The Schmitt trigger reproduces the PWM wave, but with steeper edges than when it emerges from the summing amplifier. It does this by delivering an essentially regenerated version of its input. The basic idea is that more rapid switching transitions make the class D amplifier operate more efficiently—there is less power dissipated during rise and fall times. A driver stage is included to ascertain that the power MOSFETs in the class D amplifier are turned on hard during their on periods. The low-pass filter which is inserted between the class D amplifier and the speaker load has a cutoff frequency higher than the audio spectrum, but much lower than the carrier frequency (the triangular wave). As mentioned, this filter performs integration, or *demodulation* of the PPM wave. The wave finally impressed across the speaker is a near replica of the original audio-input signal. Also shown in this diagram is an input amplifier for the audio signal. Whether this is needed for the sake of gain will depend upon the circumstances of an individual system, but it does provide an electrically convenient point to return negative feedback.

It is only fair to comment on some of the difficulties experienced with previous attempts to exploit the features of this unique approach to audio amplification. The parts count—primarily the number of ICs needed—although not excessive, could be dramatically reduced in a modern design by using one of the many dedicated IC control systems for PPM switching power supplies. These ICs incorporate all of the needed circuit functions prior to the class D output amplifier itself. Their claimed PWM frequency capability is better than 100 kHz, but accompanying charts show that this might be extended to several hundred kilohertz. (The error signal of a switching regulator would correspond to the audio input signal of the PPM audio amplifier under discussion.) That these ICs are applicable can be seen from the block diagram of a typical one shown in Fig. 5-31. They generally contain 50–100 equivalent transistors. It appears that they could greatly simplify a PWM amplifier.

It is necessary that a very high-frequency triangular wave (carrier) be used. This contributes to the ability of a simple low-pass filter to strain out the PWM wave so that the recovered audio wave is nearly free of high-frequency contamination. Previously, complementary pairs of power MOSFETs left something to be desired in the way of margins of safety. Now, however, both N-channel and P-channel types are electrically rugged and are available at consumer-oriented prices. These are used instead of bipolar transistors because it is easier to attain the switching speed needed to accommodate the several hundred kilohertz carrier switching rate.

Another reason the carrier frequency must be high is that the effective negative feedback decreases with rising audio frequency to the extent that it is difficult to obtain reasonable feedback at higher audio frequencies. Negative feedback is needed for analogous reasons to its need in conventional amplifiers—the wave processing departs somewhat from ideal performance, so negative feedback is needed to cancel the inadvertent distortion.

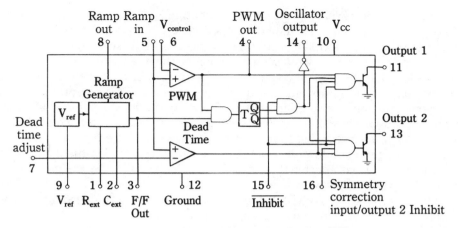

5-31   Functional block diagram of the MC3420 regulator control IC. Motorola Semiconductor Products, Inc.

Other than a suggested connection point, no feedback provision is depicted in Fig. 5-30. This is because it is felt that this remains a tricky, if not unsolved, problem. It is here that experimentation and innovation can be expected to pay high dividends. A feedback network has the same demands made on it as in conventional amplifiers, but they are more difficult to comply with in this system. Phase and amplitude of the feed-back signal are not so easy to control. If hi-fi performance is not the objective, feedback can be dispensed with.

There are other problems, too. EMI and RFI must be dealt with in a systematic manner. Among other things, this implies good RF-shielding practices. Fortunately, much has been learned about this problem from switching power supplies. From the manufacturer's point of view, it is felt that the necessary shielding, bypassing, and decoupling techniques tend to, at least partially, offset cost advantages in other areas, such as heat removal.

Although this technique may remain controversial or latent for application to hi-fi audio for the consumer's market, its advantages are often exploited where rigorous distortion specifications are not in effect.

An efficient power device for electronic ignition systems

Automobile ignition systems had a long history of maturity before electronic types gained acceptance. The performance and reliability of the erstwhile mechanical beaker system had more than proved itself in family automobiles, in racers, and in aircraft. At the same time, there had long been awareness of the limitations and shortcomings of the breaker-point actuation. Obviously, the right kind of electronic system would do away with the maintenance problems necessarily associated with the physical motion of a current-carrying switch. But, many things had to be considered before the auto makers became receptive to the newer technology.

In addition to the psychological inertia to continue in the long-comfortable way, there were matters of cost, temperature behavior, component availability, standardization, and compatibility with other devices, such as spark plugs.

The finally adopted schemes did not embrace the hottest circuits, nor the most sophisticated technology, but compromised in a way to give dramatically improved results over the older system, but at reasonable cost. This compromise meant retention of the ignition coil, the same or similar sparkplugs, and minimal upgrading of insulation. Basically, the electronic ignition system adopted by most manufacturers operates according to the same principle as did the mechanical-breaker system; the significant difference is that make and break of the primary current in the ignition coil is now accomplished electronically instead of by the movement of current-carrying contacts.

The electronic interruption of ignition-coil current is handled by a power switch such as a bipolar transistor, a Darlington transistor, an SCR, or a power MOSFET. None of these devices are ideal for the purpose; there are problems of drive, commutation, or dissipation. Also, it is desirable that the voltage drop when the switching device is ON be minimal in order to promote easy starting when the cold start of a high-compression engine causes a decrease of battery voltage. However, note that none of the devices mentioned have proven unsatisfactory because of switching speed. A worst-case analysis shows that an eight-cylinder engine turning at 6000 RPM requires 400 sparks per second. Multiplying this switching rate by ten in order to get good square-wave response indicates that the power switch need not have a response much beyond 4 kHz, a relatively low rate. Therefore, the high-frequency capability of ordinary power MOSFETs is not required.

Another shortcoming of the conventional power MOSFET for use as a power switch in ignition systems is the high drop in voltage due to high R_D that must be coped with. A quick reference to spec sheets will reveal that low R_D power MOSFETs are generally 40 to 80 V units. For ignition service, a 400 V rating is needed because that is the order of magnitude of the flyback voltage generated when the primary current of the ignition coil is interrupted. Incidentally, this is also the shortcoming of the popular power Darlington used for this purpose.

A superior power switching device is the insulated gate bipolar transistor, or IGBT. As its name suggests, this device features the easy-drive characteristic of the power MOSFET together with the high output conductivity of the bipolar power transistor. As you would suspect, there is a performance trade off suffered to achieve this desirable behavior—the switching speed is far below that of conventional power MOSFETs. Fortunately, this is of no relevance in an ignition system, where, as pointed out, a moderate audio-frequency capability suffices. Actually, the IGBT is fabricated similarly to a power MOSFET, but a modified doping profile in the drain region allows minority carriers to participate in the current-carrying process, making the operation in this region somewhat similar to that of the base-collector section of the bipolar transistor. An incidental byproduct of the modified doping is that the IGBT does not exhibit the parasitic diode common to the output sections of conventional power MOSFETs.

The basic electronic ignition system using the IGBT is shown in Fig. 5-32A. Compounding the confusion caused by the different symbols that have been used

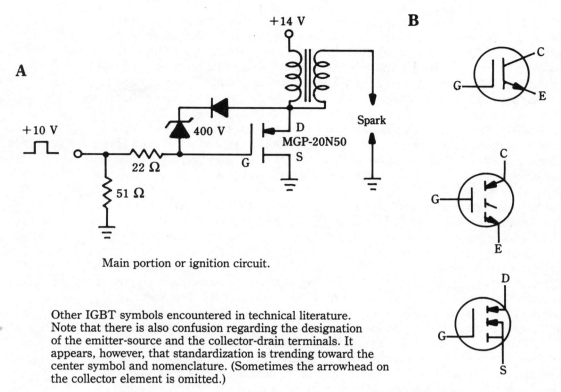

Main portion or ignition circuit.

Other IGBT symbols encountered in technical literature. Note that there is also confusion regarding the designation of the emitter-source and the collector-drain terminals. It appears, however, that standardization is trending toward the center symbol and nomenclature. (Sometimes the arrowhead on the collector element is omitted.)

5-32 Basic ignition system using the IGBT power device.

for the IGBT, are the various tradenames that have been associated with this device. Some examples are Comfet (RCA), GEMFET (Motorola), MOSIGT (IXYS Corp.), and MegaMOS (also IXYS Corp.). Also, the acronym IGT for "insulated gate transistor" is sometimes encountered. Motorola's GEMFET is particularly descriptive inasmuch as it stands for "gain-enhanced MosFET." Notwithstanding this plethora of names, the acronym IGBT generally serves to identify the device.

Early IGBTs exhibited a tendency to latch up at high currents and/or high temperatures. This has largely been overcome. IGBTs with different frequency capabilities are marketed, but those with the slower responses have the higher output conductivities (lower R_D). However, unlike conventional power MOSFETs, there is little penalty to pay in speed for high-voltage ratings.

Although not essential, greater flexibility for experimentation can result from driving the IGBT in Fig. 5-32A with a MOSFET driver IC such as the IXLD427 (made by IXYS Corp.).

Drive source for automotive fuel injector

It is easy enough to devise solenoid drive circuits for a variety of general purposes. Basically, what is needed is a suitable power device and some means of protecting it from the flyback voltage occurring when the solenoid is turned off; this precaution is necessary because of the inductance of the solenoid. Bipolar transistors, power MOSFETs, and Darlingtons have all been used successfully, as have certain power op amps. The newer IGBT devices, because of their low conductive losses, are well suited for many solenoid drive applications. Solenoids that are ac excited can be driven by SCRs or triacs, in which case the circuit is essentially that of a solid-state relay.

In spite of the straightforward approaches to many solenoid-drive applications, there are other applications where special attention is needed in order to obtain desired results. The injector solenoids in automobile fuel-injection systems is one. The emphasis here is on precise performance, efficiency, reliability, and last, but not least, low cost.

All solenoid devices tend to have some common characteristics; these can often be more or less neglected for mundane applications, but must be taken into account for such an application as the fuel-injector solenoid. Solenoid devices usually exhibit higher inductance at the leading edge of the actuating pulse because the armature is then well inside the solenoid. Solenoid devices tend to require higher initial actuating currents than holding currents. The common practice of maintaining a much higher holding current than is necessary degrades the efficiency of the system and imposes more costly design to cope with the heating problem. But even worse, the excessive current implies higher energy storage which in turn interferes with the rapidity and precision of turn-off.

From the above, note that the driving source should initially provide a high current for fast turn on, and thereafter drop the sustaining current to a lower value; also, during the sustaining or holding period, the source should act as a constant-current supply so that the actuating force on the armature of the solenoid remains constant regardless of the temperature of the windings. Yet another benefit of reducing the holding current and the stored energy is to enable the fuel-injector solenoid to operate at higher rates—clearly a basic consideration in automotive use. As might be suspected, the combination of performance requirements are best met with the use of a dedicated IC. Figures 5-33 and 5-34 depict the application and current-control feature of such an IC.

The SIDAC, a two-terminal power pulser

The SIDAC is a high-power version of the long-used triac triggering device, the diac. Like the diac, the SIDAC is a bilateral switch—it switches from its blocking to its ON state on both half cycles of the ac wave and remains conductive until some minimum value of holding current is reached. As with the diac, the SIDAC owes its switching behavior to its negative resistance characteristic. The generalized volt-ampere characteristics of diacs and SIDACs is shown in Fig. 5-35A.

A novel feature of negative-resistance devices is that they can be made to function as relaxation oscillators in an analogous fashion to a neon-bulb oscillator. The diac did not lend itself well to such use because of its limited power capability which sufficed for its intended triggering of triacs, but was inadequate for most

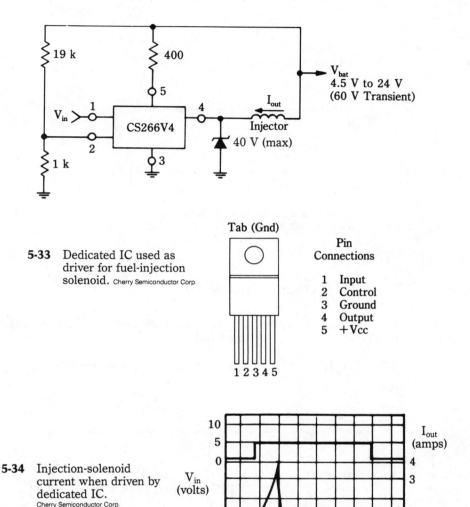

5-33 Dedicated IC used as driver for fuel-injection solenoid. Cherry Semiconductor Corp.

Tab (Gnd)

Pin Connections

1 Input
2 Control
3 Ground
4 Output
5 +Vcc

5-34 Injection-solenoid current when driven by dedicated IC. Cherry Semiconductor Corp.

other applications. On the contrary, the SIDAC, when used as a relaxation oscillator, can be put to work in gas or oil ignitors, high-voltage ignitors for discharge lamps (fluorescent, neon, flashers, lasers, etc.), CD ignition systems, electrostatic air filters, and other applications requiring bursts of high energy. The SIDAC circuits can be very cost effective because the SIDAC will often be the only active device needed.

The xenon flasher shown in Fig. 5-36 represents the basic approach to many applications depending upon oscillatory energy. Because SIDACs are available with nominal 115 or 230 V triggering levels, the dc supply voltage must be greater

A

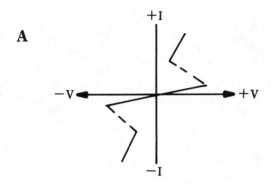

General characteristics of bilateral trigger
devices, such as the SIDAC.

B

Circuit symbol of the SIDAC.

Behavior of the SIDAC is similar to that of the diac, much used for firing thyristors,
but its voltage and current ratings are much higher.

5-35 The SIDAC—a bilateral negative-resistance trigger-device.

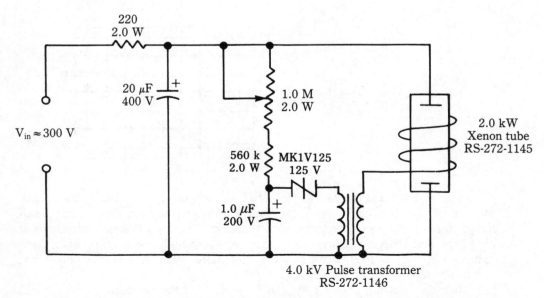

Here the SIDAC operates as a relaxation oscillator with practical power capability.

5-36 Use of the SIDAC in a xenon flasher. Motorola Semiconductor Products, Inc.

than these triggering voltages. Also, the pulse-repetition rate will be a function not only of the RC time constant of the relaxation oscillator, but also of the dc supply voltage. The circuit of Fig. 5-36 does not use the bilateral characteristic of the SIDAC, and this device does not require polarity considerations. Operation also is possible from a negative dc supply if proper consideration is given the polarized capacitors. Incidentally, certain lamps, or other loads, might exhibit polarity effects for optimum performance.

The xenon tube is self-extinguishing after being ionized for a brief instant because its sustaining current is derived from the 20 μF capacitor. As the voltage across this capacitor drops below the level needed to sustain ionization, the tube turns off. It is then retriggered by the SIDAC relaxation oscillator. Thus, a continual sequence of brief, but bright, flashes takes place.

A power-control device that has everything—almost

Previous discussions have called attention to interesting and useful modifications of the long-established power MOSFET. In one instance (see "The Nature and Implementation of the Current-Mode Power Supply" in chapter 3), an additional element was made available for electronically sensing the load current. This element, known as the *mirror, pilot,* or *sense* connection, provides a more efficient and more economical way to sense load current than the traditional source resistor, or current transformer. Motorola calls this device the *SENSEFET.*

In a second instance (see "An Efficient Power Device for Electronic Ignition Systems," in this chapter), the doping profile of the drain region is modified to allow minority, as well as majority, carriers to participate in current flow. This essentially makes the output section of the device behave much like the collector-base region of a bipolar power transistor. Appropriately, the device is often called an IGBT (insulated-gate bipolar transistor). As would be expected, the device features the high input, low-drive characteristic of a conventional power MOSFET, but the low output voltage-drop associated with bipolar power transistors.

It should not be surprising that the two modifications reviewed above have also been combined in a single device. This is somewhat reminiscent of an era of evolution of the vacuum tube in which enhanced performance was attained by adding grids and other structures. In any event, there is now the current-sensing IGT transistor pioneered by Harris Semiconductors. The symbol and packaging of this device are shown in Fig. 5-37. The features of this device can be indeed remarkable. It can carry voltage and current ratings of 500 V and 50 A. Note that conventional power MOSFETs or bipolar power transistors with such ratings would need large metallic packages, such as the TO-3 or stud-mounting provisions. The small TO-218 package suffices for this device because of its inordinately low conductive losses. This comes about through its very low $V_{CE(sat)}$ of about 1.5 V at 50 A. Or, thinking of it as a power MOSFET, its R_{DS} is approximately 52 milliohms.

There is a trade off for such outstanding performance parameters. Its turn-off speed is relatively low, so that high switching losses limit its use to about 5 kHz. It

TERMINAL DIAGRAM

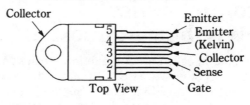

92GS-44006R1

N-CHANNEL ENHANCEMENT MODE

TERMINAL DESIGNATION

Collector

5
4
3
2
1

Emitter
Emitter
(Kelvin)
Collector
Sense
Gate

Top View

TO-218 (5 LEAD)

Besides providing electronic sensing of load current, such a device features high voltage and high current ratings, and easy-drive requirement.

5-37 Symbol and pinout of a current-sensing IGBT.

remains eminently suited to many power control applications involving ac and dc motors, power supplies, and drivers for solenoids, contactors, and relays.

The overcurrent protection scheme shown in Fig. 5-38 can be used in a variety of applications. It makes use of the current sensing portion of the device to produce the requisite disable signal when excessive load current exists. The current sensed across the 1 k ohm resistor is proportional to the collector-emitter current and therefore to the load current. Because of the minute current provided by the sensing element (here labeled P for *pilot*), power dissipation in the 1 k resistor is negligible. The load in this arrangement could represent a motor, heater, solenoid, or the output circuit of a power supply. Turn off of the load current occurs when both inputs of the NAND gate are no longer high.

An overload protection such as this is often desirable because this power device can deliver very high current under fault conditions. This not only endangers the load, but it is possible for the device to latch-up in its ON state. Although such latch-up is the result of grossly abusive conditions, it is common for certain loads, especially motors, to occasionally encounter conditions demanding much greater than normal currents. When appropriately used, the IGBT-type power devices are destined to compete seriously with bipolar power transistors,

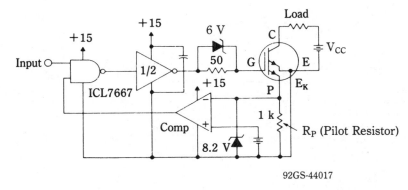

92GS-44017

The IGBT can provide very high load current. Also, excessive overloads can produce a latched-on condition of the device. Therefore, it is expedient to make use of the current-sensing feature for limiting load current.

5-38 Basic overcurrent protection technique for current-sensing IGBTs.

Darlingtons, power MOSFETs, and thyristors in many applications. In particular, loads requiring tens of amperes at several hundred to perhaps, one thousand volts, but at switching rates below 5 kHz merit consideration for IGBT-based design. As pointed out, the power device just discussed is basically an IGBT with the added refinement of a current-sensing element.

The MCT—a near-ultimate power device of the latching family

The current-sensing IGBT just described as a power-control device with almost everything belongs to the class of devices exemplified by bipolar transistors, power Darlingtons, and power MOSFETs. All of these devices are nonlatching—load current is always under full control of the input element. Full control implies that the load current can be smoothly varied from zero to maximum by varying the amplitude of the input signal.

In contrast, there are power-control devices known as thyristors; these are latching devices, exemplified by the SCR and the triac. Once these devices are triggered into conduction, the input element (gate) loses control over the load current. Unique among this class of power-control devices has been the GTO (gate turn off) thyristor. Unlike the conventional SCR, load-current in the GTO can be turned off (the device can be unlatched) by application of a reverse-polarity pulse to the gate circuit. The general idea for implementation of the GTO is shown in Fig. 5-39. Note that there is no commutating circuitry even though operation is from a dc source. Excellent efficiency is attainable—a six A GTO switched at 20 kHz with a 50-percent duty cycle can handle a kilowatt load at 95 percent efficiency. Nonetheless, the popularity of the GTO has seemingly been less than

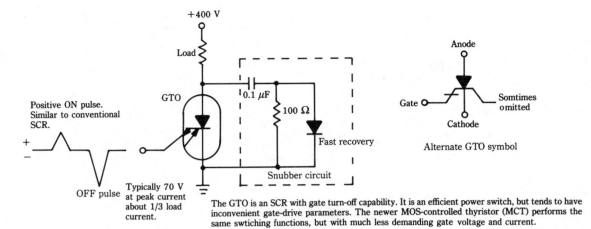

Positive ON pulse. Similar to conventional SCR.

OFF pulse

Typically 70 V at peak current about 1/3 load current.

The GTO is an SCR with gate turn-off capability. It is an efficient power switch, but tends to have inconvenient gate-drive parameters. The newer MOS-controlled thyristor (MCT) performs the same swtiching functions, but with much less demanding gate voltage and current.

5-39   Driven inverter using the GTO thyristor.

deserved. One reason certainly stems from the inconveniently high negative voltage and peak current required for turn off; negative 70 V, and about 1/3 load current are typical turn-off requirements.

What would be needed is a latching device with MOSFET gate so that the low voltage drop and high load-current capability of SCRs could be retained along with a very low power turn-on and turn-off feature. Such a device would avoid the gate-drive shortcoming of the GTO. From a practical standpoint, such a hybrid component would indeed be the near-ultimate power-control device for applications where latching behavior is acceptable.

As might be expected, such a power-control device does, indeed, exist. It is the MCT (MOS-controlled thyristor). As its name suggests, its gate is the easy-drive input of MOSFETs; its output is a triple-junction structure that, like an SCR, regeneratively latches ON during the conductive state. Unlike the conventional SCR, this device readily unlatches on reversal of gate polarity. Unlike the GTO, negligible peak power is needed for the turn-off pulse.

Actually, two MOSFETs are integrated into the structure of the device, an N-channel type for turn-off by a positive gate signal, and a P-channel type for turn-on by a negative gate signal. The gates of the MOSFETs are tied together so that the MCT has a single gate terminal. Turn-off occurs by breaking the feedback path responsible for the latch condition.

The MCT, as thus far described, qualifies as a uniquely useful power-control device by virtue of its controllability. However, it exhibits yet another outstanding feature—the current density of its output section greatly exceeds all other power-control devices, both latching and nonlatching. In this respect, it more closely resembles the IGBT in basic architecture. In any event, it is possible to produce a doping profile that allows exceedingly high current capability from a given-size die. At the same time, the voltage drop is comparable to that of a conventional

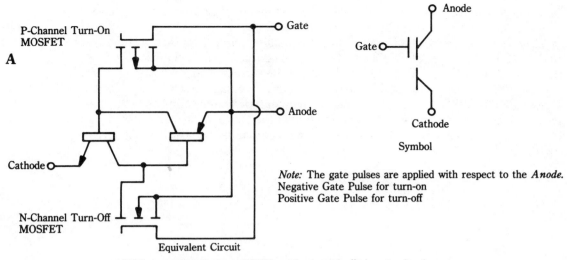

A

P-Channel Turn-On MOSFET

N-Channel Turn-Off MOSFET

Cathode

Gate

Anode

Equivalent Circuit

Anode

Gate

Cathode

Symbol

Note: The gate pulses are applied with respect to the *Anode*.
Negative Gate Pulse for turn-on
Positive Gate Pulse for turn-off

MOS-controlled thyristor (MCT) can be turned off via gate signal.

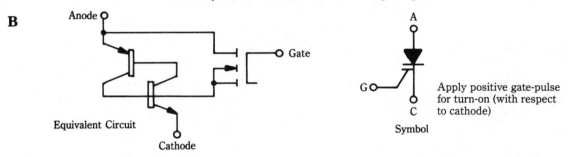

B

Anode

Gate

Cathode

Equivalent Circuit

A

G

C

Symbol

Apply positive gate-pulse for turn-on (with respect to cathode)

MOS thyristor cannot be turned off by gate signal. This device, once popular, is an endangered species, having been superseded by the more-versatile MCT.

5-40 Similarly named but different thyristors using field-effect gates.

SCR. Experimental units have been investigated that can turn off 100 A, 1000 V loads in 2 microseconds. Faster turn-off times will be forthcoming in future development, and a turn-on time of 200 nanoseconds is already attainable.

The MCT or MOS-controlled thyristor is not to be confused with a similarly named device, the MOS thyristor, which appears to be fading from the scene. The MOS thyristor was somewhat similar to the MCT, but contained only a single MOSFET in its structure. Because of this, it was easily turned on, but could not be turned off by a gate signal. It was, in other words, similar to a conventional SCR in this respect. Accordingly, it could not be used in a dc circuit such as shown for the GTO inverter in Fig. 5-39. Some means of commutation would have to be provided, because once turned on, the gate loses control and the device remains on as long as its anode-cathode holding current is above a certain minimum value. Although the MOS thyristor is destined to be superceded by the more-controllable MCT, enough

MOS thyristors were used to justify mention. They did, after all, feature an easy-drive characteristic compared to conventional SCRs.

Figure 5-40 shows the difference in the equivalent circuits of the two devices. In the MCT, one of the MOSFETs serves to turn on the device in similar fashion to the single MOSFET in the MOS thyristor. But, because a P-channel MOSFET is used for this purpose in the MCT, the required turn-on pulse is negative in polarity. The second MOSFET in the MCT is an N-channel type and accomplishes turn-off by shorting the feedback path causing latchup of the two bipolar transistors. Note that the turn-off pulse impressed at the common gate must be positive in polarity.

Index

Prices Subject to Change Without Notice.

Look for These and Other TAB Books at Your Local Bookstore

To Order Call Toll Free 1-800-822-8158
(in PA, AK, and Canada call 717-794-2191)

or write to TAB BOOKS, Blue Ridge Summit, PA 17294-0840.

Title	Product No.	Quantity	Price

☐ Check or money order made payable to TAB BOOKS

Charge my ☐ VISA ☐ MasterCard ☐ American Express

Acct. No. _____ Exp. _____

Signature: _____

Name: _____

Address: _____

City: _____

State: _____ Zip: _____

Subtotal $ _____

Postage and Handling
($3.00 in U.S., $5.00 outside U.S.) $ _____

Add applicable state and local
sales tax $ _____

TOTAL $ _____

TAB BOOKS catalog free with purchase; otherwise send $1.00 in check or money order and receive $1.00 credit on your next purchase.

Orders outside U.S. must pay with international money order in U.S. dollars.

TAB Guarantee: If for any reason you are not satisfied with the book(s) you order, simply return it (them) within 15 days and receive a full refund.
                                                                    BC

THE MASTER IC COOKBOOK
—2nd Edition—Clayton L. Hallmark and Delton T. Horn

"A wide range of popular experimenter/hobbyist linear ICs is given in this encyclopedic book."
—Popular Electronics, on the first edition

Find complete information on memories, audio amplifiers, RF amplifiers, and related devices, in addition to other sections on TTL, CMOS, special-purpose CMOS, and other linear devices. Circuits come complete with pinouts, specifications, and a concise description of the IC and its applications. 576 pages, 390 illustrations. **Book No. 3550, $22.95 paperback, $34.95 hardcover**

49 EASY ELECTRONIC PROJECTS FOR THE 747 DUAL OP AMP—Delton T. Horn

Explore the powerful 747 dual op amp as you experiment with these practical applications for this inexpensive, widely available, and easy-to-use workhorse circuit. You will find completely tested instructions for all the projects, including operational circuits, audio projects, signal generators, filters, test equipment, modulation projects, pulse circuits, and miscellaneous projects. 190 pages, 146 illustrations. **Book No. 3458, $15.95 paperback, $23.95 hardcover**

49 EASY ELECTRONIC PROJECTS FOR TRANSCONDUCTANCE & NORTON OP AMPS—Delton T. Horn

The projects cover a wide range of practical applications from dc amplifiers to current switches, voltage regulators to Schmitt triggers, and more. Each includes easy-to-follow instructions, and most can be constructed in a single evening costing less than $15 to build. Delton T. Horn gives you all the information you need to use transconductance and Norton op amps in your projects. 230 pages, 163 illustrations. **Book No. 3455, $16.95 paperback, $25.95 hardcover**

49 EASY ELECTRONIC PROJECTS FOR THE 556 DUAL TIMER—Delton T. Horn

Perfect for beginning to intermediate electronics experimenters, this project book features 49 projects designed around the 556 dual timer. The 556 dual-timer IC contains two independent 555-type timers in a single package, making many sophisticated applications possible. Simple, step-by-step building instructions for each application and many detailed drawings, diagrams, and schematics make tackling these projects easy. 190 pages, 130 illustrations. **Book No. 3454, $15.95 paperback, $23.95 hardcover**

WRITING BETTER TECHNICAL ARTICLES—Harley Bjelland

Publishing technical articles is a sure bet to enhance your professional career. With this unique new stylebook, you can develop the writing and editing skills needed to get published! This guide leads you through all the steps between the idea and the sale. The author targets the shorter technical article, but his techniques are equally well suited to all types of technical writing, including books, manuals, proposals, and letters. 208 pages, 40 illustrations. **Book No. 3439, $12.95 paperback, $19.95 hardcover**

THE SECRET LIFE OF QUANTA
—Dr. M. Y. Han—Foreword by Eugen Merzbacher, Ph.D.

Now your questions about high-tech physics are answered in plain, low-tech English! When you try to find out exactly *why* atoms do what they do, you'll find that many of the texts are written for physicists, not for the reader who is simply curious. That's what makes this such an important publishing milestone. This is your handbook for the 21st century! **198 pages, 103 illustrations. Book No. 3397, $17.95 hardcover only**

Other Bestsellers of Related Interest

HANDBOOK OF DATA COMMUNICATIONS AND COMPUTER NETWORKS—2nd Edition —Dimitris N. Chorafas

Completely revised and updated, this results-oriented reference—with over 125 illustrations—progresses smoothly as theory is combined with concrete examples to show you how to successfully manage a dynamic information system. You'll find applications-oriented material on networks, technological advances, telecommunications technology, protocols, network design, messages and transactions, software's role, and network maintenance. 448 pages, 129 illustrations. **Book No. 3690, $44.95 hardcover only**

DISASTER RECOVERY HANDBOOK —Chantico Publishing Company, Inc.

Could your company survive if a tornado struck today? You'll find everything you need for coping with your worst-case scenario in this book. Among the other issues covered are plan formulation and maintenance; data, communications, and microcomputer recovery procedures; emergency procedures. Action-oriented checklists and worksheets are included to help you start planning right away—before it's too late. 276 pages, 88 illustrations. **Book No. 3663, $39.95 hardcover only**

BUILD YOUR OWN MACINTOSH® AND SAVE A BUNDLE—Bob Brant

Building your own Mac is easy and much less expensive than buying one off the shelf. Just assemble economical, readily available components for your own catalog-part Macintosh, or "Cat Mac," with the help of this book. You can also upgrade a Mac you already own, add more memory, a hard disk, a bigger display, or accelerator card to your existing systems—all at the best price. 240 pages, 113 illustrations. **Book No. 3656, $17.95 paperback only**

BUILD YOUR OWN 80486 PC AND SAVE A BUNDLE—Aubrey Pilgrim

With inexpensive third-party components and clear, step-by-step photos and assembly instructions—and without any soldering, wiring, or electronic test instruments—you can assemble a 486. This book discusses boards, monitors, hard drives, cables, printers, faxes, modems, UPSs, memory, floppy disks, and more. It includes parts lists, mail order addresses, safety precautions, troubleshooting tips, and a glossary of terms. 240 pages, 62 illustrations. **Book No. 3628, $16.95 paperback, $26.95 hardcover**